67th Porcelain Enamel Institute Technical Forum

67th Porcelain Enamel Institute Technical Forum

Proceedings of the 67th Porcelain Enamel Institute Technical Forum, Nashville, Tennessee, USA (2005)

Conference Director
Steve Kilczewski

Assistant Conference Director
Holger Evele

Editor
William D. Faust

Published by

The American Ceramic Society
735 Ceramic Place
Suite 100
Westerville, Ohio 43081
www.ceramics.org

67th Porcelain Enamel Institute Technical Forum

For information on ordering titles published by The American Ceramic Society, or to request a publications catalog, please call 614-794-5890, or visit www.ceramics.org.

ISSN 0196-6219

ISBN 1-57498-278-8

10 09 08 07 06 5 4 3 2 1

Contents

Preface

The entire Technical Forum Committee is pleased to deliver to you the proceedings of the 67th Annual PEI Technical Forum. This volume represents the successful completion of a year's worth of planning and preparation, culminating in three days of meetings and seminars at the Doubletree Downtown Hotel in Nashville, Tennessee on May 2-5, 2005. As you receive these proceedings, work is already progressing on the 68th Technical Forum, to be held on May 15-18, 2006 in Nashville, Tennessee, again this year at the Doubletree Downtown Hotel, Nashville.

With the close of the 67th PEI Technical Forum, I have completed my two-year term as Chairman of the Technical Forum Committee. It was my honor to lead such a fine group of dedicated individuals. Foremost, I would like to thank my vice-chairman Holger Evele (Ferro), as well as the members of the Technical Forum Committee for their time, efforts, and supportive endeavors on behalf of this year's forum. The success of the forum is directly attributable to their contributions. Each year we strive to uphold the tradition of offering information that has both useful and practical applications for our industry. I believe that we have accomplished that challenge, and have set the bar even higher for upcoming years.

The Back-to-Basics Seminar continues to be an important element of the Technical Forum. Attendance at the B2B continues to remain stable. Our gratitude goes to Holger Evele for orchestrating another outstanding seminar, and to the faculty staff that implemented the program along with Holger. This seminar continues to be a well-attended favorite, attracting both newcomers to the porcelain enamel industry, as well as seasoned veterans. Again, thanks to all involved.

The thanks of the entire Committee goes out to this year's excellent group of speakers who provided us with information on the latest in materials and equipment used in the porcelain enameling process. We are grateful to them for their time and efforts in researching, preparing and presenting their informative papers. We are very appreciative to Mr. Bob Long for presenting us with "The Six Decades of Porcelain Enameling." Further thanks go to those suppliers who participated and supported the ever-popular Supplier's Mart. Our final thanks go to all who attended and participated in the 67th Annual PEI Technical Forum.

Holger has assumed the chairmanship and will be heading up the planning efforts for the 2006 Technical Forum. Please welcome Mr. Peter Vodak (Engineered Storage Products) as the new vice-chairman for the years of 2006 and 2007. With the dedication and guidance of these two fine gentlemen, we can look ahead to the future with assurance that the Technical Forum will continue to highlight the best that our industry has to offer. Please mark your calendar for next year's meeting May15-18, 2006 in Nashville. It promises to be instructive and worthwhile.

Steve Kilczewski, Pemco Corporation
Chairman 2005 PEI Technical Forum Committee

SIX DECADES OF PORCELAIN ENAMELING

Robert Long
American Porcelain Enamel Company

Abstract
Robert Long provided an overview of six decades of enameling art and science reflecting in his broad experiences and active involvement in the industry. Early in his career he attended Muskegon Community College in Muskegon, Michigan. In 1995 Robert Long chaired the 17th International Enamellers Congress in Nashville, Tennessee.

Introduction

Robert Long started in the enameling industry at the age of 14 when he visited some customers with his father, the founder of American Porcelain Enamel in Muskegon, Michigan. His first job was spraying stipple on various cooker parts. His active career after military service started in the late 1940's. At this time, significant changes in the industry were occurring as the US and other countries were converting from war materiel production to consumer products again.

Changes in the Enameling Industry

One of the aspects of enameling at this time was the use of open hearth furnaces for manufacturing ingot cast steels. The cause of fishscaling was not known and steel companies felt it was an enamel fault. The ground coats were generally high fire, about 1600°F with the steels having a very strong alkaline cleaning and acid etch followed by a borax-soda ash neutralizer. The acid of choices for etching the steels were either sulfuric or muriatic.

Application of the enamels was by dipping, usually manual but gradually supplanted by various automated systems moving parts through the dipping tanks. The effects of the borax-soda ash neutralized on enamel dipping properties such as pick-up and drain time were part of the "art" of enameling controlled by the individuals skilled in the process. One interesting aspect of art was the "beading" of the edges of cookware. The beading was done by skilled individuals using their hands to put a very precise and narrow band of a soft glass of the edges of the steel. This banding helped to terminate the thick enamel edge with a high expansion glass to minimize edge stresses. The bands were usually bright red, black or blue.

Various shapes of kitchen utensils coated with enamel by slushing or dipping process.

Cookware with edge "beading" applied by hand.

Enamels were nearly always milled by the end user in a variety of ball mills. The "mill room" was a key part of the enameling plant and the individuals managing the work were highly skilled in manipulating the mill formulations to achieve optimum application.

A modern mill room, showing enamel grinding mills, scales, enamel containers, and two small color-matching mills in the foreground.

Typical mill room of the first half of the 20th century for preparation of wet enamels.

Cast iron enameling was more widespread than it is today. Casting were coated with "white" ground coats which were then dried and fired if a wet process was used or directly if a dry process was used. Subsequently, either clear or opacified enamel was applied as a finish coat. The opacifier of choice was antimony oxide smelted into a glass and required 10 or more mils of application for satisfactory coverage. This was said to be the most difficult type of product to run in an enameling plant. In the 1930's through the late 1970's, many of the dry process cast iron product producers were "self-smelters". Self-smelters produce their own frit products in-house and are integrated through the finished product. Self-smelters still represent significant tonnage of frit produced in the US.

The types of cover coats available before 1940 were either antimony or zircon opacified. These coatings had good alkali resistance but poor acid resistance. Acid resistance was achieved by a clear overcoat. These coatings had an interesting degree of apparent depth but also were very thick with a total thickness approaching 20 mils. Brushing of the edges of various parts was normally done to minimize edge spalling or chipping. Appliances of this era are easily identified by the black edges seen on component parts. In the 1950's, with the advent of better sources of titanium dioxide and an understanding of what caused yellowing of the titanium oxide in glasses, high reflectivity acid resistant enamels were now possible. Today nearly all of the bright white porcelain enamels are titanium dioxide opacified.

The advent of nickel pickle, nickel sulfate baths for replacement deposition of nickel metal on the steel surfaces after acid etching, allowed the development of "high speed ground coats". The high speed ground coats are similar to the standard ground coats that are used today, but in the early 1950's, this allowed the development of ground coat enamels firing about 1450°F. At the same time, high opacity titanium cover coats firing about 1500°F provided the enameller with greatly improved products. The titanium opacified cover coats were two-thirds the weight of the older zircon and antimony enamels. Additional development of the pickling systems resulted in the development of high etch and high nickel deposition allowing "direct-on" enameling with only the cover coat. This improved the mechanical properties of the finish coats with less chipping tendency and the elimination of any brushing.

Batch pickling operation moving baskets of parts from various tanks with cleaner, acid, nickel and neutralizer solutions to prepare the metal surface for application of enamel.

Low temperature enameling of about 1200°F to 1350°F seemed feasible. However, due to the cost of lower temperature enamels with more costly raw materials and the loss of radiation heat transfer below 1350°F, temperatures of enameling settled to about 1500°F. About this time, it was determined that the cause of fishscaling was due to the lack of micro voids in the steels.

This spurred on the steel suppliers to produce fishscale resistant steel typically termed low carbon steels. Micro voids in the steel were created by the development of iron carbides, which are crushed during cold rolling and then eliminated by open coil annealing (OCA). In the early 1970's, various continuously cast steels were developed and adapted to the enameling processes. At the same time, enamel products began the development of no-nickel, no-pickle ground coats, which eliminated the previously necessary pickling process of acid etching and nickel deposition. This greatly reduced the solid waste associated with the pickling processes.

The application of wet spraying techniques and equipment processed from the early 1930's to highly automated systems in the 1960's with equipment suppliers providing better spraying equipment. The traditional dipping operations were used primarily of complex and bulky shapes such as refrigerator liners and dishwasher interiors. This was also automated with wet spray and robotic application in the early 1970's.

Robotic application of cover coat enamel, direct-on to dishwasher cavities.

Summary

Continuing developments such as electrostatic powder has helped the industry compete with organic coatings. The mentality of our industry is and has been, "we can do more." We have kept abreast of the ever-changing process and regulatory environments and kept enameling a premier coating technology recognized for its leading attributes.

QUALITY IMPROVEMENT TECHNIQUES

Alan Woodward
Electrolux Home Products

This paper focuses on tools and techniques utilized to maintain process consistency. Daily and shift-to-shift inconsistencies in the performance of a two-coat, one-fire process was the driving force to establish methods employed by each shift on a daily basis.

Prior to the establishment of these tools it was not unusual to leave for the day with the system running with exceptional first pass yields, only to return the next day to a system generating more rework than acceptable product. Rejects such as light and heavy spray were predominant. Upon investigation it was discovered that the spray equipment had been altered, and setups changed.

The challenge became quickly obvious. That was to standardize the set ups used between all shifts to a common baseline. By doing so the process can be "centered" in the process window.

Gun Position – Identified optimum gun position for each product to be run. This included absolute position, oscillator distance, oscillator height, gun to target distance, and gun angle. Tools needed are a tape measure, angle gauge, and a plumb line. A standard set up sheet was created which defined these criteria. Each shift then became responsible for verifying that all guns were in proper position and certifying as such by signing off. If changes to gun position do become necessary, there is an area for notations so changes can be documented and followed.

Gun Settings – The optimal voltage, flow rates, and atomizing air were addressed in a similar manner using standardized set up sheets by product family.

Environmental Conditions – Relative humidity and spray room temperature were always defined but frequently ignored. In order to make booth operators aware of their environment, control charts were put in place. By plotting temperature and relative humidity on an hourly basis the operator became intimately aware of their spray room conditions. These charts were banded into three color-coded ranges – green for normal operating range, yellow for correction needed, and red for the need to shut down the process.

In conclusion for consistent success, it is critical to *"Control the Process Inputs not react to process outputs."*

Quality Improvement Techniques - Key Factors Identified

- Gun Position
- Distance To Target
- Gun Angle
- Gun Voltage
- Flow Rate
- Atomizing Air
- Relative Humidity

FOUNDRY PROCESS CONTROLS FOR PORCELAIN ENAMELING CAST IRON

Liam O'Byrne
O'Byrne Consulting Services

Abstract
Factors needing to be controlled in the foundry casting process to ensure maximum chance of successfully enameling the solidified casting will be briefly discussed. Possible effects of not controlling the factors will be explained.

Introduction
Last year, a paper discussing the preferred chemical composition of cast iron for porcelain enamel was presented. It was indicated that having the right composition is only the beginning of the story for producing castings suitable for enameling. This paper will briefly discuss some of the processing parameters that the foundryman must control in order to ensure the best possible substrate for subsequent porcelain enamel coatings.

The key parameters the foundryman must consider are:

- Chemical Composition
- Molten Metal Processes
- Sand System Control

The success of any enamel operation on the castings produced by the foundry will ultimately depend on how well the foundryman accounts for variations in these parameters during the casting process.

Chemical Composition
By way of review, the standard ranges for the main elemental components of cast iron for enameling were presented as follows:

Total Carbon	3.2% – 3.6%	
Graphitic Carbon	2.8% - 3.2%	
Combined Carbon	0.2% - 0.5%	
Silicon	2.3% - 2.7%	
Phosphorus	0.3% - 1.0%	
Sulfur	0.05% - 0.1%	
Manganese	0.4% - 0.7%	(S%x1.7) + 0.3%
Carbon Equivalent	4.2% - 4.6%	TC% + (Si% + P%)

Controlling these elements within their suggested ranges offers the best possibility for producing a metal structure that will be suitable for porcelain enamel coatings.

The ideal structure for porcelain enameling is a precipitate of randomly oriented primary graphite flakes in a ferrite or ferrite/pearlite matrix. Pearlite is a lamellar matrix of Ferrite and Iron Carbide (Fe_3C). While pearlite is a harder phase than ferrite (pure iron), it is still adequate for enameling providing a blast cleaning process is used that will properly etch the casting surface.

The foundryman's challenge is to develop the correct structure by controlling the other key parameters in the casting process.

Molten Metal Processes

The Melting Furnace

The first control point the foundryman must consider is the equipment used for melting his metal composition. Two basic types of melting furnace are used today:

1. Cupola

The cupola is basically a steel cylinder charged with coke, limestone, pig iron and cast iron scrap. Air is blasted through a series of openings in the cupola wall called tuyeres, and the air is used to provide the melting energy. A good mixing of chemical constituents is an advantageous part of this melting process, and the temperature of the molten metal leaving the cupola can be controlled closely by the air blast through the tuyeres. Additional chemical inoculation for compositional and structural control can be used when filling the casting mold.

2. Electric Furnace

The electric furnace is charged with limestone, scrap and pig iron. The main types of melting furnaces are the induction furnace and electrical resistance furnace, although in the cast iron foundry, induction furnaces are generally used.

A coreless induction furnace can charge and melt raw, solid scrap, while a conventional induction furnace will allow solid scrap additions, but must keep a molten "heel" of metal circulating in the inductor.

Some mixing does occur during the melting process, although it is generally not as efficient as in the cupola. Again, chemical inoculation is used when needed for compositional and structural adjustment when filling the mold.

Cooling Rates and Pouring Temperature

Providing a homogeneous molten metal raw material is only the beginning for the foundryman. Significant structural variations can be produced by the actual casting process itself. Primary Graphite is the goal for the microstructure of the casting, and this is promoted by slower cooling in the mold, and having preferred "nucleation" sites for the graphite in the molten liquid. Slow cooling rates are not easily achieved in most castings for enameling due to the generally thin casting sections. Raising pouring temperatures to properly fill the thin sections tends to dissolve any nucleation sites in the molten metal. If the foundryman tries lowering pouring temperature to prevent this, it can cause problems with the mold not filling properly, leading to cold shuts and misruns. The end result is that the foundryman has to perform a "balancing act" with the process to arrive at the best compromise.

Inoculation

Fortunately, the foundryman does have help in developing the best possible microstructure, using the process known as inoculation. Primary Graphite is the goal, but fast cooling rates in thin sections tend to prevent primary graphite forming. The needed help for primary graphite phase solidification is by providing "nucleation sites" for graphite precipitation. These sites require less overall energy for primary graphite solidification.

Inoculant (usually a ferrosilicon alloy) addition provides microscopic nucleation sights in the molten metal for elemental carbon to grab hold of. This helps achieve primary graphite precipitation with a random orientation during fast cooling rates. The graphite precipitates along with solidifying pure metal "dendrites". These metal dendrites have a "Christmas tree" structure. As cooling proceeds, the inoculant effect fades, and graphite precipitation is only achieved along preferred orientation planes between the randomly solidified metal dendrites.

It is unlikely however, that all carbon will precipitate out as graphite, due to the fast cooling rate. Some carbon will remain combined in the metal matrix as Fe_3C (cementite). It will be present either in a Pearlite phase mixed with ferrite, or as a pure Cementite (chilled iron) phase, especially at or near the casting surface. This is a very hard metallic phase, which is difficult to blast clean properly, so again the result is a balancing act for the foundryman.

Sand System Controls

Basic Green Sand Constituents

The basic constituents of most green sand systems used for the vast majority of cast iron molds are as follows:

- Synthetic Sand or Natural Sand
- Facing Sand (if required)
- Bonding Clay (Southern/Western Bentonite)
- Water
- Coal Dust (or other organic constituent)

Sand

The choice of synthetic or natural sand must be the first decision. A synthetic sand is basically silica grains with clay, water and organic additives. A natural sand contains its own clay bonding material so generally only needs water and organics additions. It may have a facing sand used at the metal/mold interface. Generally, synthetic sands are used in high pressure sand molding operations, while natural sands are used in lower pressure molding systems.

The sand grain size and shape must be chosen for smoothest surface while providing adequate venting for gas escape during pouring. Vents are generally added to mold patterns to help with gas escape through the sand mold.

An incorrect base sand choice and preparation can lead to many different defects; expansion defects, (rat tails, scabs, etc.); pin hole defects; sand burn on and metal penetration.

Bonding Clay (Southern and Western Bentonite)

The bonding clay used in a mold to hold the sand grains together must be strong enough to maintain mold strength during pouring and solidification/cooling. The minimum amount required to achieve the mold strength and dimensional stability needed should be used.

The right mix of Southern and Western Bentonites should be used for the correct wet and dry strengths in the mold. Southern Bentonite is a calcium-based clay and tends to have higher wet strength in the mold and lower dry strength when the hot metal has been poured into the mold. Western Bentonite is a sodium based clay, and has higher dry strength and lower wet strength. The clays are added to the sand and mixed in a machine called a "Mullor;" hence the mixing operation is called "Mulling." Adequate mixing (mulling) time should be taken for the maximum sand strength development with minimum clay content.

As the clay content increases in the sand, the water demand of the sand increases for a specific level of strength development. Incorrect clay levels and/or mulling procedures can cause mold collapse, clay ball defects, expansion defects, gas and sand pin holing.

Water Content

Water in the sand mix works in conjunction with the clay to develop bond and mold strength. The water makes the clay "swell" and encapsulate the sand grains. The clay becomes plastic, and during the mulling process, deforms around the sand grains, coating them. As clay content increases, the water demand increases for adequate bond development. Too little water creates "friable" brittle sand, causing sand defects such as pinholes, sand wash, and mold collapse. Too much water causes excessive gas (steam) evolution causing sand burn on, scabs, and possibly even mold explosions. Excess water can also cause deep, chilled iron at the casting surface, due to faster cooling rates in the mold. The high specific heat of water removes thermal energy very quickly from the metal at the metal-mold interface.

Organics
Organic additions are used primarily to produce a smoother surface on the casting.
Coal dust, cellulose or other types of organic material burn quickly in the mold, producing a thin gas layer between the metal and sand interface. The gas layer prevents excessive metal penetration of the mold surface, helping to provide a smoother surface than would otherwise be achieved.

Too little organic material in the sand mix causes excessive metal penetration, sand burn on and rough casting surface. Too much organic material can cause gas defects, carbon buildup on the surface of the casting with consequent blister defects in enameling, and porosity. A bluish tint may also be seen on the surface, if high levels of coal-based organic materials are present.

Conclusion
The Foundrymans Dilemma
Many process parameters in the foundry operation can act against each other in terms of the types of casting defects they produce or prevent. Some of these control parameters are additive in nature, in that they can interact together to increase the overall tendency for a particular defect.
Others are mutually exclusive and work in opposite directions to each other. The foundryman is called upon to perform a "Balancing Act" with his processes for producing the best possible casting. In many cases, there is no substitute for past experience when developing a foundry process control system for a new casting for enameling.

ADVANCEMENTS IN AUTOMATIC SPRAY APPLICATION OF PORCELAIN ENAMEL

Mitchell Drozd
ITW Industrial Finishing

Abstract

Equipment specifically designed for aqueous suspensions of ceramic glazes and porcelain enamels is described. Wet application spray guns have been used for many decades in the porcelain enameling industry worldwide. Adaptation to stationary, pneumatic gun movers, robots and other automated systems has provided a high level of application efficiency.

However, the abrasive nature of glazes, porcelain enamels, inorganic components such as quartz, alumina and feldspar and the need to maintain the spray equipment at high levels of operational efficiency has provided opportunities for improvement. The use of cemented tungsten carbide components in wet application spray equipment combined with improved and simplified designs is providing 4 to 7 times the life of components compared to conventional alloys used in spray equipment.

New designs are simpler, with fewer parts allowing for faster and easier maintenance. The new designs also allow better cleaning which reduces corrosion of components. Quick mounting is also a feature of the newer design. Overall, operating costs are lowered with this newer equipment.

Presentation Overview

- Coating
- Application Equipment
- Typical Products (Substrates) Coated
- Economic Drivers in Spray Application - ROI
- Spray Gun Innovations
- Conclusions

Coating

- **Porcelain Enamel :**

Inorganic coating applied to a substrate and fused at temperature above 800° F.

Also referred to as vitreous (glassy) enamel

Coating

Coating Mixtures Typically Consisting of :

- **Clays** - is a generic term for an aggregate of hydrous ***silicate particles*** less than 4 µm (micrometres) in diameter
- **Talcs** - A whitish, greenish, or grayish hydrated magnesium ***silicate mineral*** that is extremely soft and has a characteristic soapy or greasy feel.
- **Flint** - (or flintstone) is a hard, sedimentary cryptocrystalline ***silica rock*** with a glassy appearance. Flint is usually dark grey, blue, black, or deep brown in color. It occurs chiefly as nodules and masses in chalks and limestones.
- **Proprietary Compounds** - to add texture, color, surface finish, etc.
- **Water**

Application Equipment

Typical Fluid Application Equipment:

- Tumblers - coatings ingredients mixed
- Holding vats – coating seasoned
- Tote tanks – coating ready to spray
- Pumps – coating delivered to spray station or between tanks (High Volume Production)
- Pressure pots – coating delivered to spray device (typically lower volume production)
- ***Spray Guns – automatic spray application device***

Application Equipment

Typical Application Machinery:

- Stationary Guns
- Pneumatic Gun Movers
- Reciprocating Machines
- Robots
- Chain on edge conveyors
- Overhead Conveyors

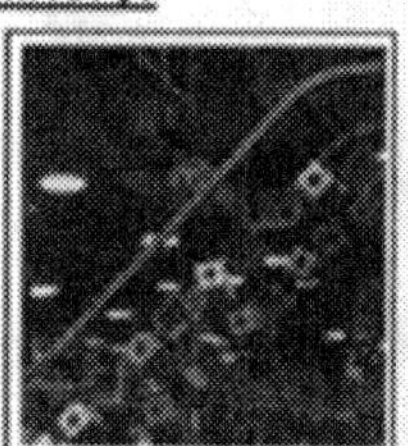

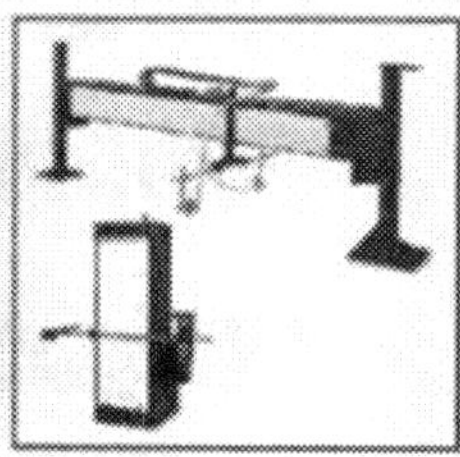

Typical Products Coated

- Cooking Appliances
- Tableware
- Sanitary Fixtures
- Floor and Wall Tiles
- Electric Insulators
- Home Appliances

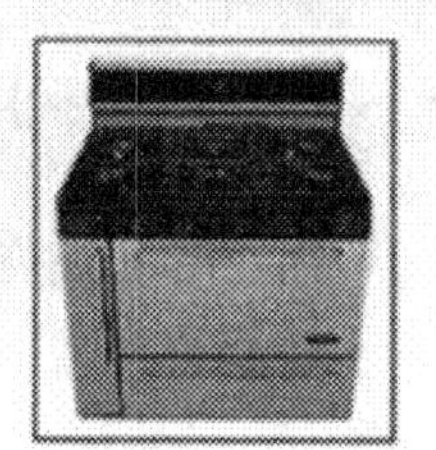

Typical Application - Summary

Operating Environment:

- Abrasion
- Equipment contamination
- Overspray / Dust
- High Humidity

Problems to deal with:

- Accelerated wear
- Cleaning
- Filtration / Contamination
- Corrosion

Bottom Line: High cost of maintenance and replacement parts !!

Economic Drivers in Spray Application

Spray Gun Performance Considerations

- Atomization – coating finish quality
- Air consumption – cost of compressed air
- Durability – wear, metallurgical considerations
- Maintenance – Down time, speed of repair
- Production requirements –defects, versatility

Economic Drivers in Spray Application

Return on Investment Analysis (ROI) Issues to consider when purchasing an -Automatic Spray Gun-

- Fluid Nozzle Costs
- Fluid Needle Costs
- Atomizing Air Cap Costs
- Number of Spray Guns
- Durability of all above (how long do they last ?)
- Labor Rate to service guns
- Time it takes to service guns

Automatic Spray Gun Innovations

Ceramic Coating Spray Gun

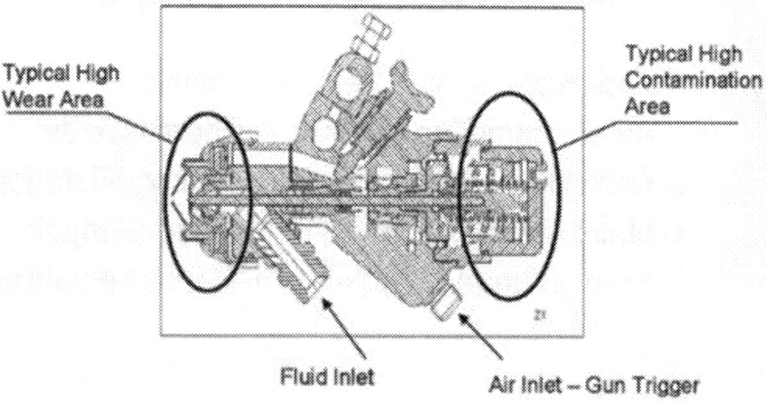

Typical High Wear Area

Typical High Contamination Area

Fluid Inlet

Air Inlet – Gun Trigger

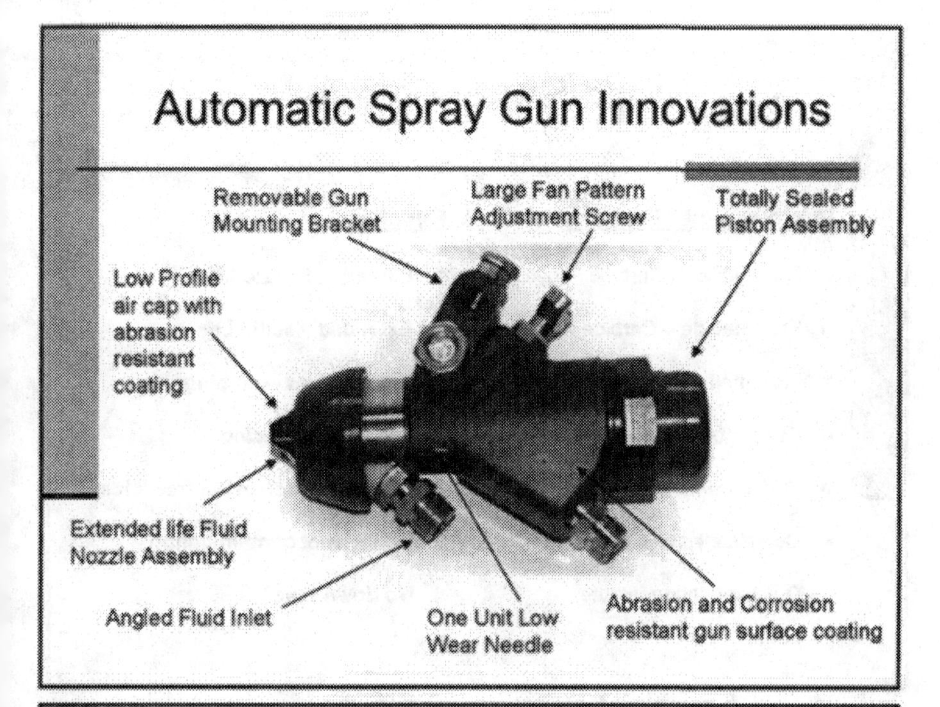

Automatic Spray Gun Innovations

- ASTM Abrasion Test methods conducted to assess abrasion resistance of key spray gun parts
- Typically Cemented Tungsten Carbide Alloys with Cobalt additives prolong life of gun parts extensively
- Tungsten Carbide Alloy parts assembled into high abrasion areas of spray guns
- Stainless Steel Alloys utilized in flow through areas of spray guns

Automatic Spray Gun Innovations

Improvement - Features	**Benefits:**
Fluid Nozzle - Carbide Alloys	**Extended Nozzle Life**
Fluid Needle – Carbide Alloys	**Extended Needle Life**
Flow method through the gun	**Smooth flow – no plugging**
Fewer components	**Minimum servicing**
Wash down surface coating	**Keeps gun corrosion free /clean**
Completely sealed	**No internal contamination**
Quick mounting feature	**No down time**

Conclusions

- Porcelain enamels are very difficult spray gun application environments
- High Maintenance areas in the plant
- Downtime is costly to production

Conclusions

- With proper testing and alloy selection methods, typical life expectancy is 4:1 to 7:1 time longer vs. conventional alloys
- Fewer gun parts make spray guns easy to maintain
- Superior gun surface materials protect internal and external components from wear and corrosion
- ROI on the spray gun purchase is shorter
- **BOTTOM LINE – LOW MAINTENANCE, TROUBLE FREE SPRAY GUN OPERATION = $ SAVINGS!!.**

ENERGY MARKET OVERVIEW AND OUTLOOK

Brian Habacivch
Constellation Energy, Baltimore, Maryland

Abstract

Constellation Energy is a national distributor of gas and electricity. An overview of the natural gas and electric markets from 1991 to 2005 is presented. Historical data on natural gas and electrical energy usage and supply are illustrated.

The demand for natural gas in the USA is increasing about 2% to 4% annually while production has increased at 0% to 2% annually. The gap is being filled by imports from Canada as natural gas and liquefied natural gas (LNG) from offshore. The number of wells producing gas in the USA has not kept pace with the output of wells which is declining.

Electrical energy demand is increasing for industrial, commercial and residential users. Higher prices for fossil fuels (gas, coal, oil) are inevitable. Energy reliability is emerging as a key concern. This is stimulating new interest in nuclear energy generation and renewable energy sources. In the near and mid-term there will be upward pressure on all energy prices domestically and globally.

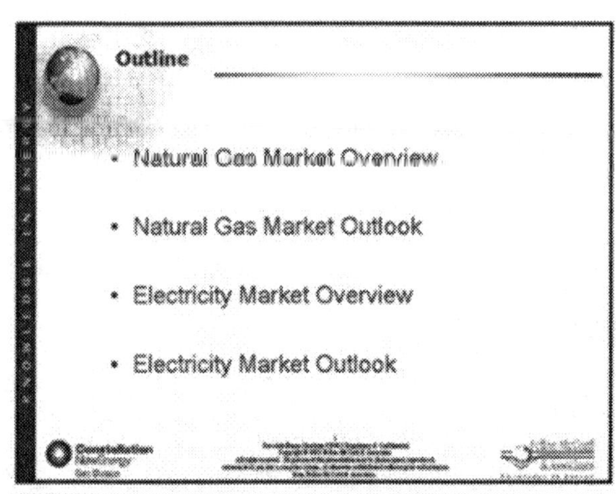
Outline
• Natural Gas Market Overview
• Natural Gas Market Outlook
• Electricity Market Overview
• Electricity Market Outlook

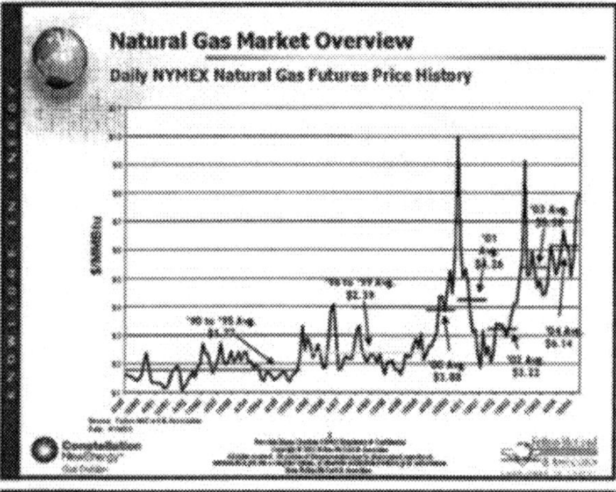
Natural Gas Market Overview
Daily NYMEX Natural Gas Futures Price History

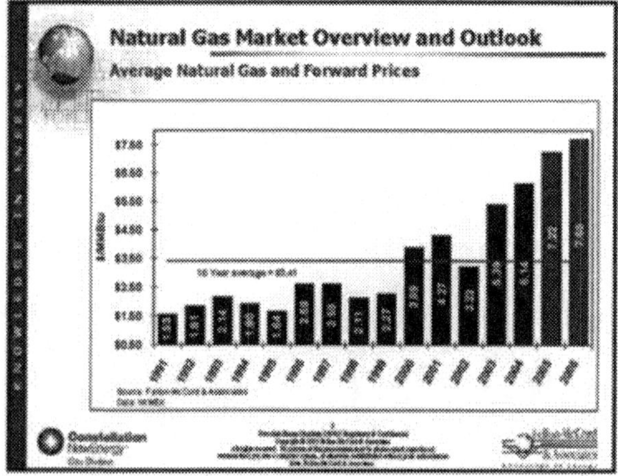
Natural Gas Market Overview and Outlook
Average Natural Gas and Forward Prices

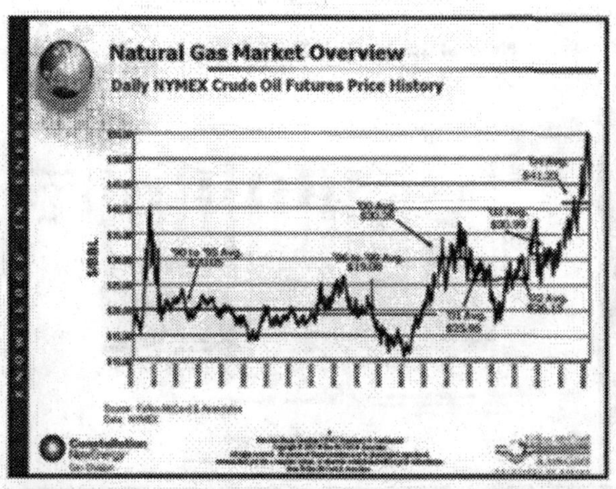
Natural Gas Market Overview
Daily NYMEX Crude Oil Futures Price History
$/BBL

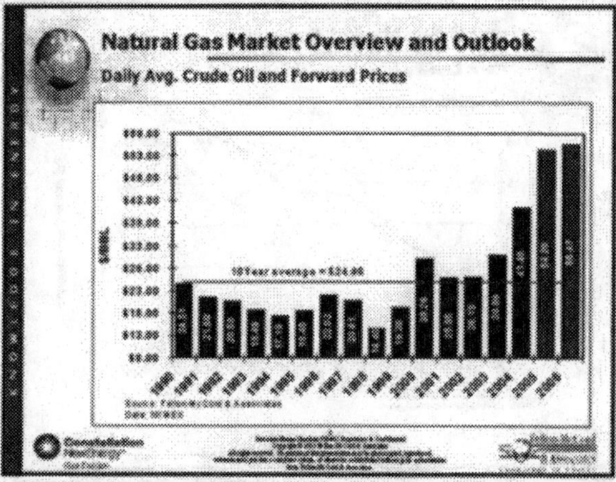
Natural Gas Market Overview and Outlook
Daily Avg. Crude Oil and Forward Prices
$/BBL

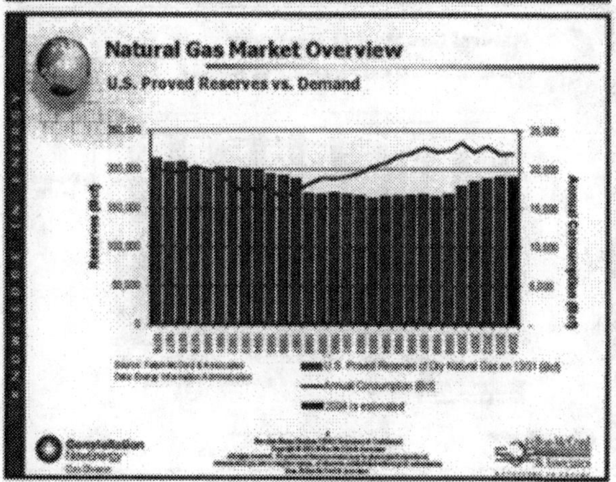
Natural Gas Market Overview
U.S. Proved Reserves vs. Demand
250,000
200,000
150,000
100,000
50,000
0
25,000
20,000
15,000
10,000
5,000
Annual Consumption (Bcf)
2004 is estimated

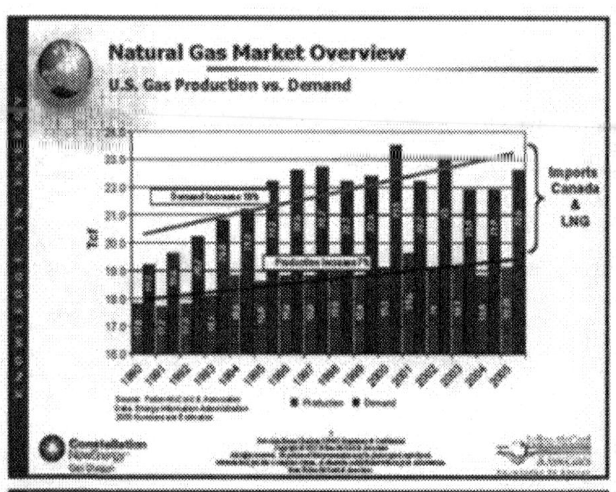
Natural Gas Market Overview
U.S. Gas Production vs. Demand
Tcf
Imports Canada & LNG

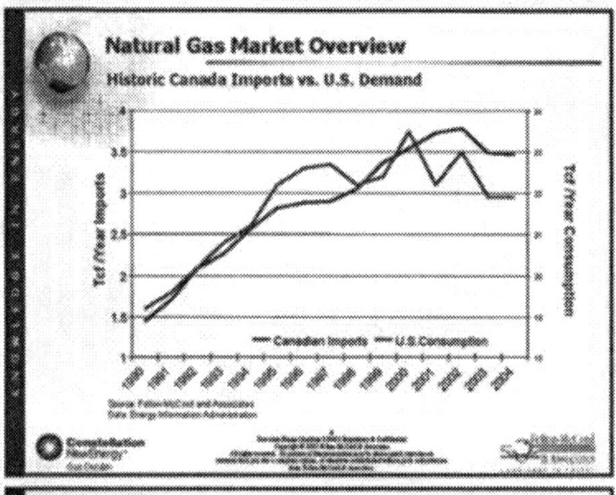
Natural Gas Market Overview
Historic Canada Imports vs. U.S. Demand
Tcf /Year Imports
Tcf /Year Consumption
Canadian Imports
U.S.Consumption

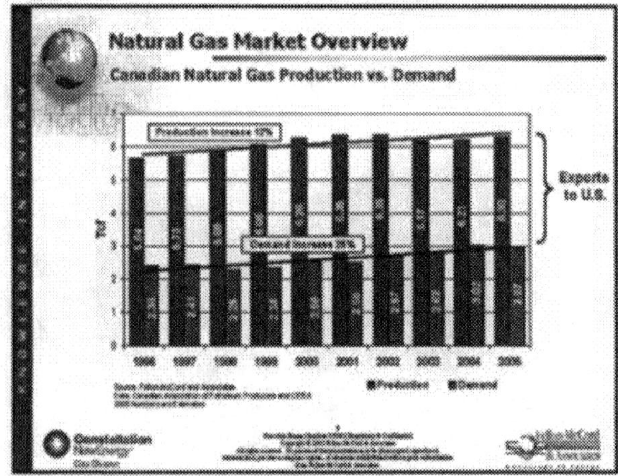
Natural Gas Market Overview
Canadian Natural Gas Production vs. Demand
Tcf
Exports to U.S.

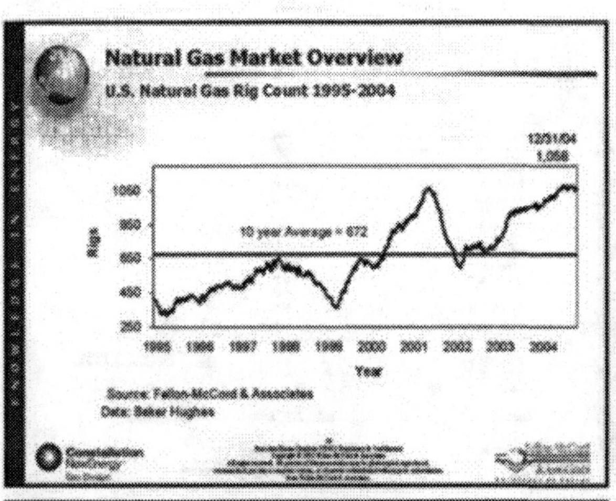

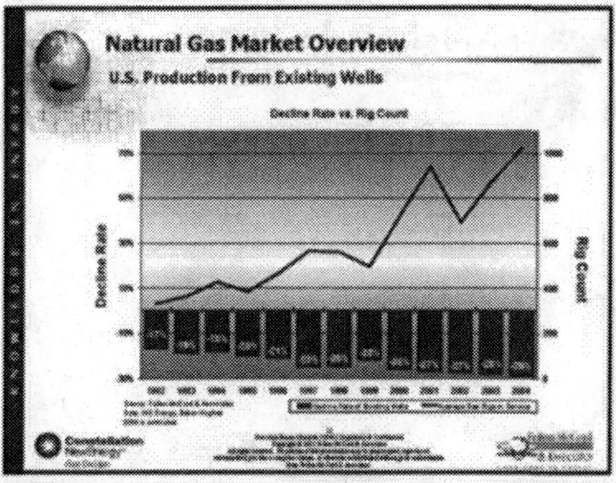

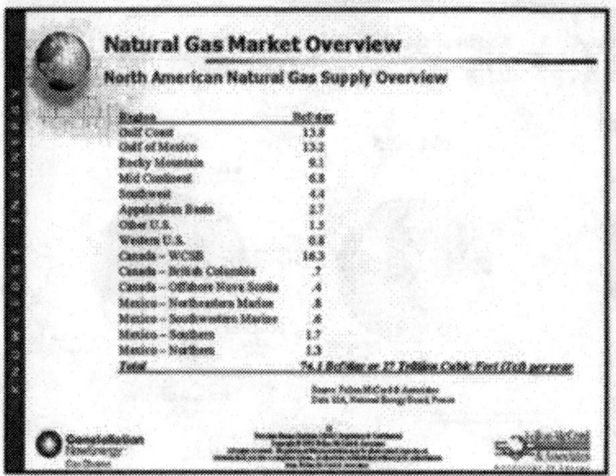

Region	Bcf/day
Gulf Coast	13.8
Gulf of Mexico	13.2
Rocky Mountain	9.1
Mid Continent	6.8
Southwest	4.4
Appalachian Basin	2.7
Other U.S.	1.5
Western U.S.	0.8
Canada – WCSB	16.3
Canada – British Columbia	.7
Canada – Offshore Nova Scotia	.4
Mexico – Northeastern Marine	.8
Mexico – Southwestern Marine	.6
Mexico – Southern	1.7
Mexico – Northern	1.3
Total	*74.1 Bcf/day or 27 Trillion Cubic Feet (Tcf) per year*

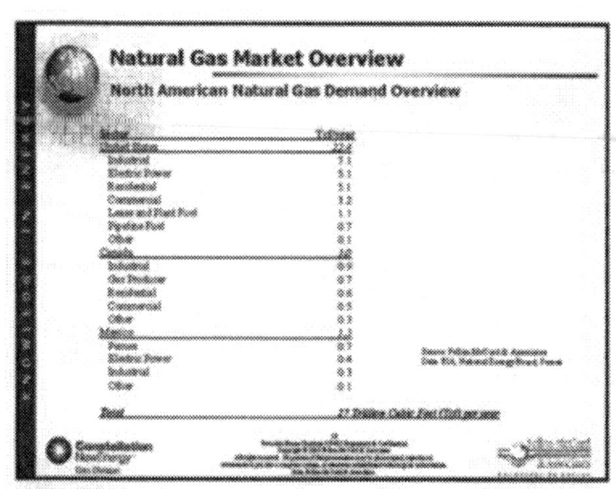
Natural Gas Market Overview
North American Natural Gas Demand Overview

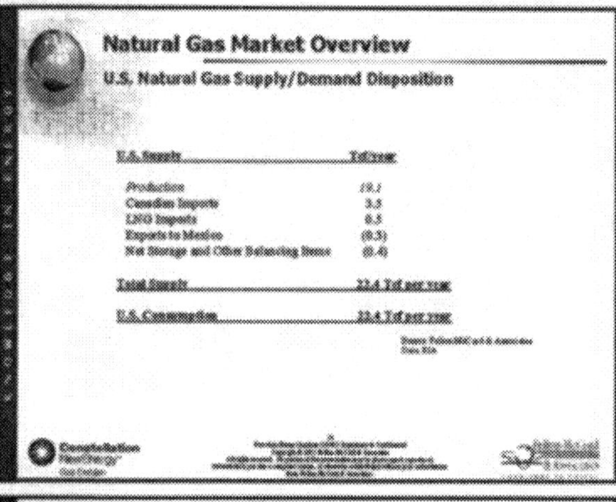
Natural Gas Market Overview
U.S. Natural Gas Supply/Demand Disposition
Production
Canadian Imports
LNG Imports
Exports to Mexico
Net Storage and Other Balancing Items
Total Supply
U.S. Consumption

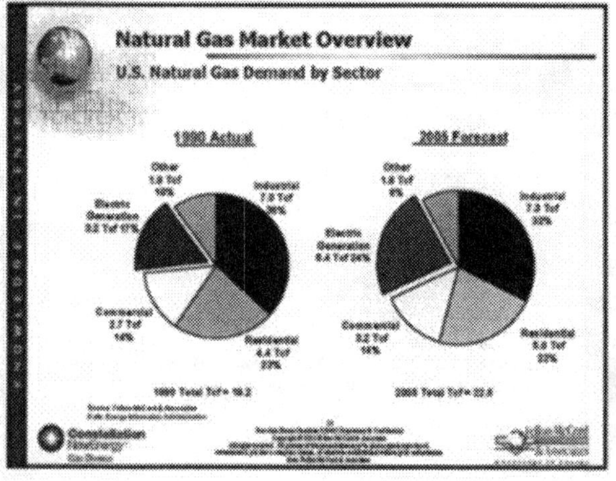
Natural Gas Market Overview
U.S. Natural Gas Demand by Sector
Other
Industrial
Electric Generation
Commercial
Residential
Other
Industrial
Electric Generation
Commercial
Residential

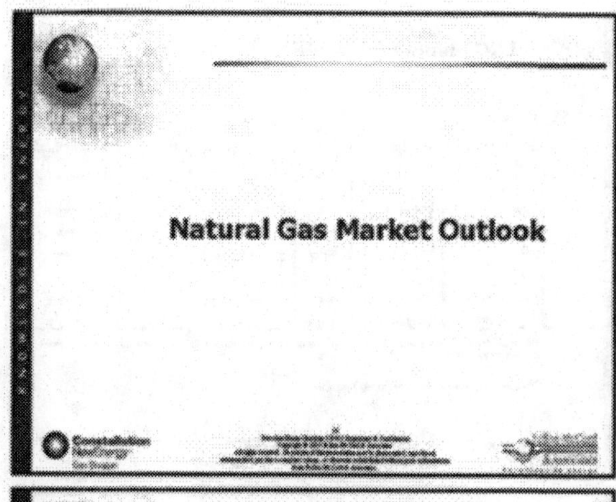
Natural Gas Market Outlook

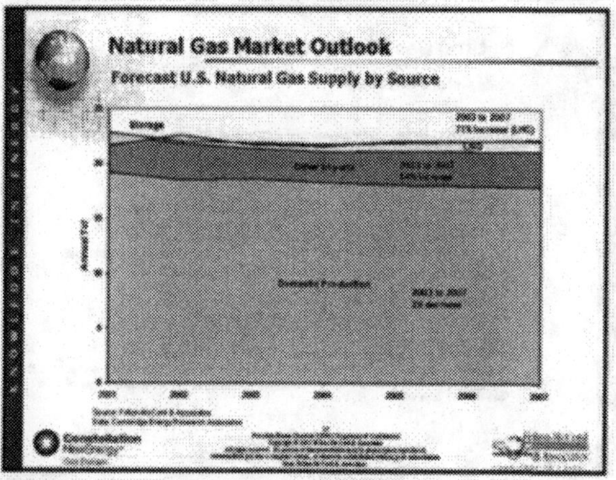
Natural Gas Market Outlook
Forecast U.S. Natural Gas Supply by Source

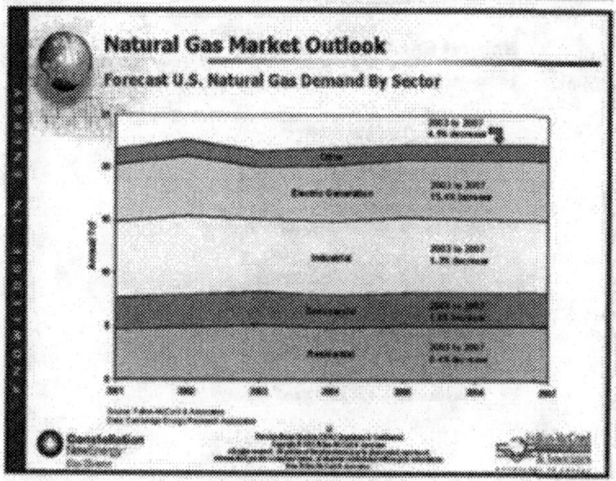
Natural Gas Market Outlook
Forecast U.S. Natural Gas Demand By Sector

LNG Summary Update

Developer/Project	Location	Max Capacity	Supply	Construction Start Date	Est. Completion Date
[illegible]	Offshore Gulf of Mexico	[illegible]	[illegible]	[illegible]	[illegible]
[illegible]	Freeport, TX	1.5 Bcf/D	ConocoPhillips [illegible]	[illegible]	[illegible]
[illegible]	Sabine Pass, LA	[illegible]	------	[illegible]	[illegible]
[illegible]	[illegible]	[illegible]	[illegible]	[illegible]	[illegible]
[illegible]	[illegible]	[illegible]	[illegible]	Under Construction	[illegible]
[illegible]	Gulf of Mexico	[illegible]	[illegible]	------	[illegible]
[illegible]	[illegible]	[illegible]	[illegible]	------	[illegible]

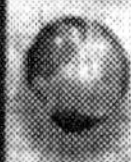

Natural Gas Market Outlook

Fellon-McCord's View

- Natural gas production expected growth of 0% to 2% per year
- Natural gas demand expected growth of 2% to 4% per year; outpacing all other energy sources
- Continued high prices and high price volatility for 2005
- Continued tight supply/demand balance through 2005
  - 2005 North American supply essentially flat to 2004
  - 2005 North American demand expected to increase 1-3% year-on-year versus 2004
- Tight supply/demand balance for next 3 years
- Extreme price volatility for the next three years
- Little supply side relief with respect to gas prices in 2005
- LNG is a longer term potential solution but imports increase only marginally 2005 versus 2004 with potential for measurable benefit in 2008 and beyond

Natural Gas Market Outlook

Fellon-McCord's View (continued)

2005 Upside Gas Price Scenario

- Early hot start to summer
- Spike in crude oil prices
- Continued weak/falling U.S. dollar
- Active hurricane season
- Strong growth in U.S./China/India GDP

2005 Downside Gas Price Scenario

- Cool summer (i.e. past two summers)
- Softening crude oil prices
- Strengthening U.S. dollar
- Softening U.S./China/India GDP
- Less than expected hurricane activity

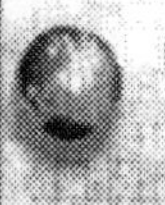

Natural Gas Market Outlook

Fellon-McCord's Recommendation

- Set quarterly goals for 2006 hedge position
  - Particular attention to Q1 2006
- Extend diminishing hedge position into 2007 and 2008

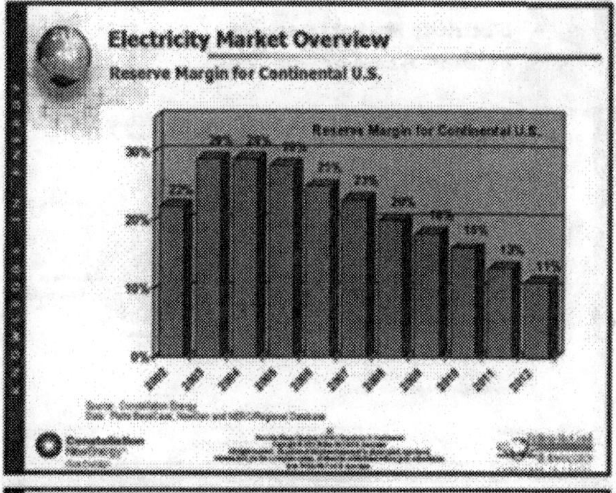

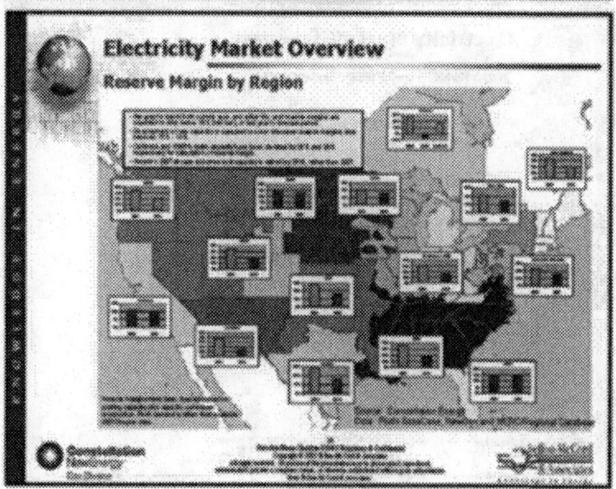

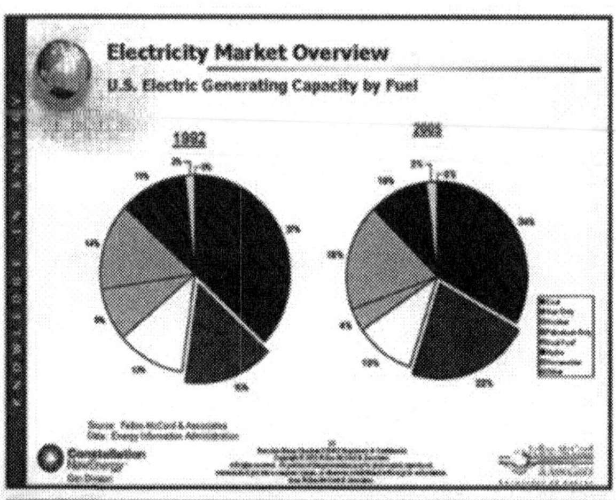
Electricity Market Overview
U.S. Electric Generating Capacity by Fuel
1992
Constellation NewEnergy

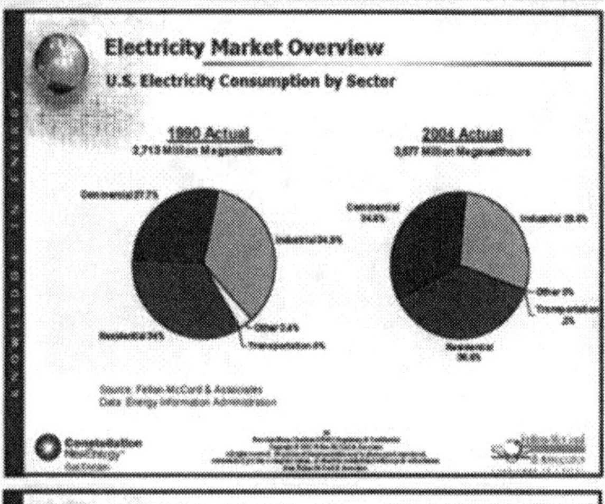
Electricity Market Overview
U.S. Electricity Consumption by Sector
1990 Actual
2004 Actual
Constellation NewEnergy

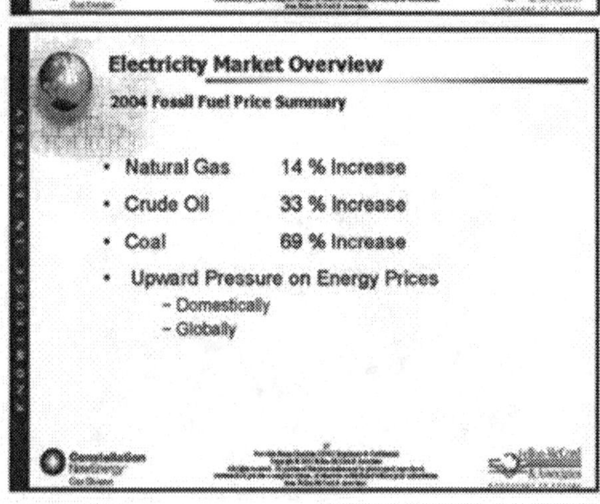
Electricity Market Overview
2004 Fossil Fuel Price Summary
• Natural Gas 14 % Increase
• Crude Oil 33 % Increase
• Coal 69 % Increase
• Upward Pressure on Energy Prices
– Domestically
– Globally
Constellation NewEnergy

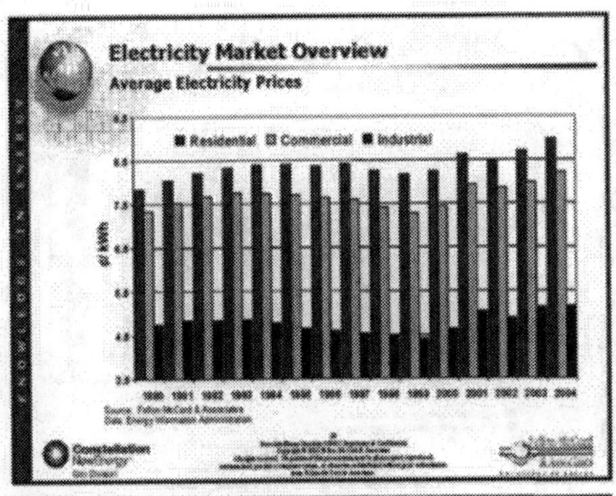
Electricity Market Overview
Average Electricity Prices
Residential
Commercial
Industrial

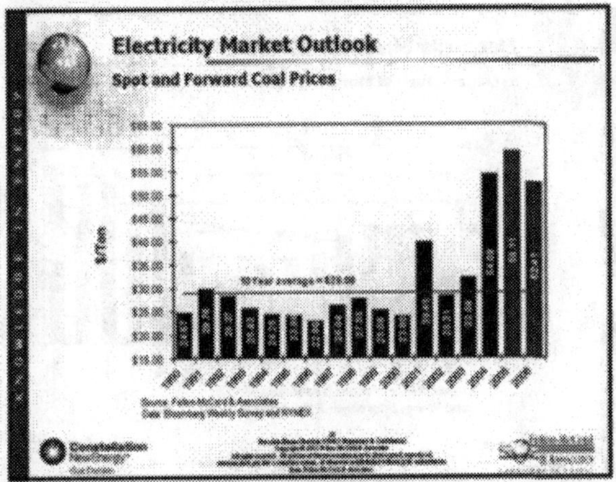
Electricity Market Outlook
Spot and Forward Coal Prices
$/Ton

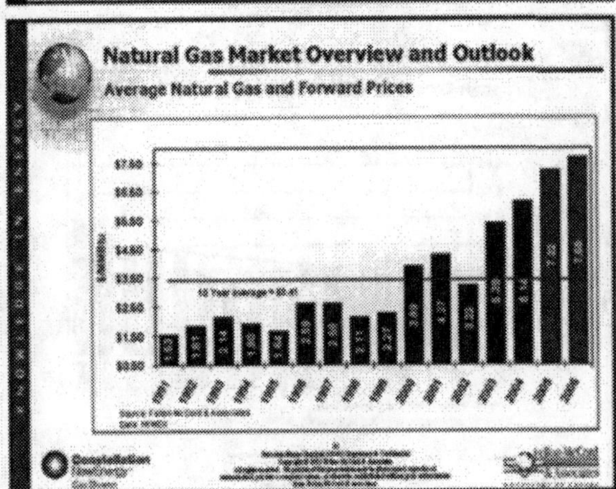
Natural Gas Market Overview and Outlook
Average Natural Gas and Forward Prices

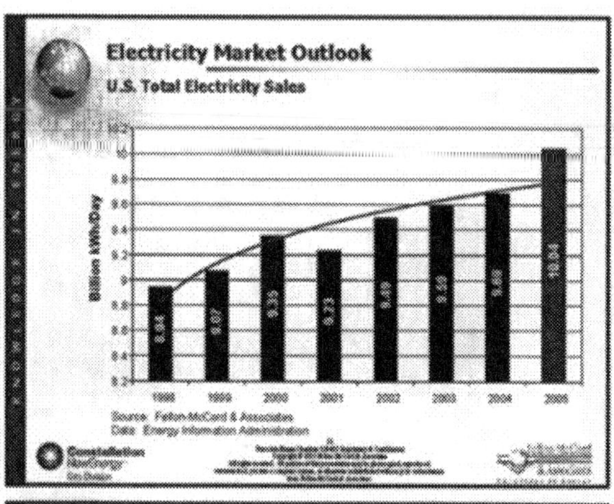
Electricity Market Outlook
U.S. Total Electricity Sales
Billion kWh/Day
Source: Fellon-McCord & Associates
Data: Energy Information Administration

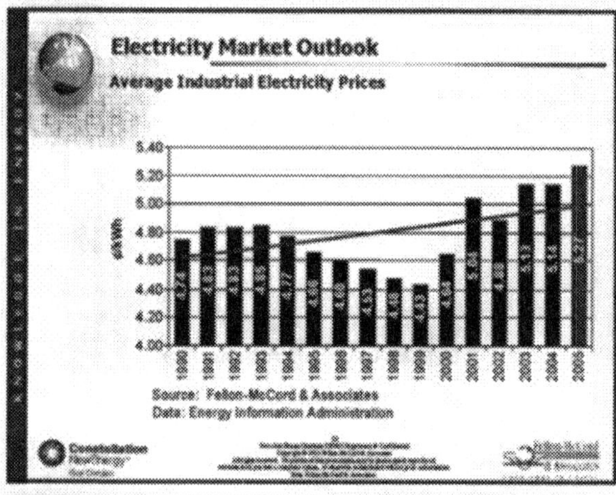
Electricity Market Outlook
Average Industrial Electricity Prices
5.40
5.20
5.00
4.80
4.60
4.40
4.20
4.00
¢/kWh
Source: Fellon-McCord & Associates
Data: Energy Information Administration

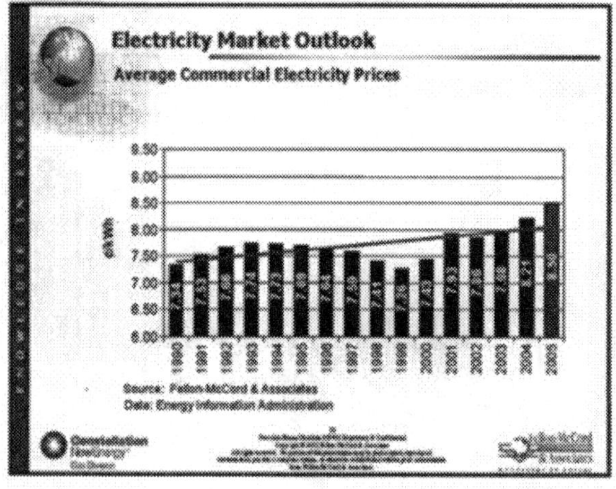
Electricity Market Outlook
Average Commercial Electricity Prices
8.50
8.00
7.50
7.00
6.50
6.00
¢/kWh
Source: Fellon-McCord & Associates
Data: Energy Information Administration

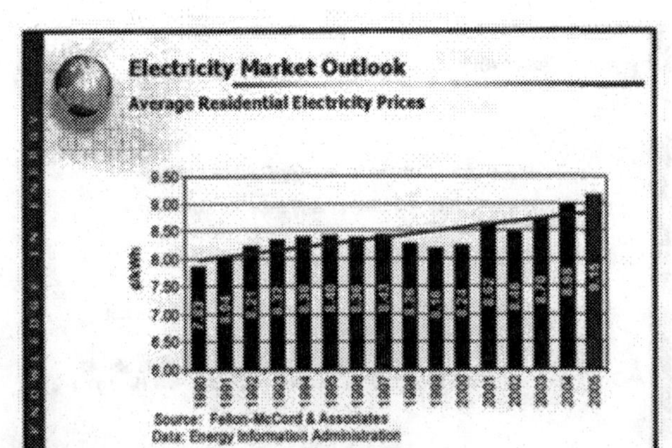

Electricity Market Outlook

Fellon-McCord's View

- Electricity Reserve Margins Peaked 2004-2005 – Declining through 2012
- Fossil Fuel Prices High Relative to Past 15 Years
- Legacy Low-Price Coal Contracts for Many Utilities Expiring in Near Term
- 90% of All New Generation Built During Past 15 Years is Gas Fired
- Emphasis on Clean Coal, Renewables, and Possible Focus on New Nuclear Technologies
- Grid Reliability to be Major Focus of Regulators/Utilities – Increase in Capital Costs
- Upward Pressure on Electricity Prices, Near and Mid-Term

Electricity Market Outlook

Fellon-McCord's Recommendation

- Ontario – Preparing to shut down 7,000 MW coal-fired power by 2007
  - 25% of Ontario's load
  - Higher-priced alternatives
  - Reliability issues
- Recommendation – If not a member become active in Canadian Manufacturers & Exporters. Contact Ian Howcroft at 905-568-8300, ext. 256
- Recommendation – FMA to abstract Hanson electricity supply across all N.A. facilities (special project basis); rank order by load, load factor, and regulatory opportunity

Summary

- North American Demand for Energy Continues to Grow
- Access to Readily Available Energy Supplies Significantly Diminished
- Fossil Fuels Face Upward Price Pressure Over Next 3-5 Years
- Return to "Cheap" or "Easily Sourced" Energy Supplies Not Likely in Near and Mid-Term
- In Addition to Enhanced Energy Efficiency, Consumers Should Focus on Price Predictability, Security of Supply, and Managing Volatility

Disclaimer

The data and information contained in this presentation are gathered and provided to Fellon-McCord & Associates through proprietary and public sources and are published with the intention of being accurate. Fellon-McCord & Associates, its parent company and any affiliates cannot, however, insure against or be held responsible for inaccuracies and futher assumes no liability whatsoever arising from use of such data or any information contained in this presentation.

The material in this presentation does not, in any way, represent a recommendation of any kind that you or your company purchase or sell any commodity. Discussions or representations of past market performance do not predict future market results. Any forecast of potential future energy prices or market trends are for discussion purposes only and are expressly not intended to induce the purchase of any commodity of any kind.

THE STRENGTH OF STEEL; THE BEAUTY OF GLASS

Warren Norton
Dofasco Steel, Hamilton, Ontario, Canada

Abstract
A review of Primary and Finishing steelmaking facilities is provided. Production of vitreous enameling steel via open coil annealing (OCA) and interstitial free (IF) processes are compared. Quality characteristics of both OCA and IF steels pertaining to press and enamel shops are discussed.

Introduction
The combination of glass on steel makes a beautiful and strong materials system. For the porcelain enamel coating to bond and create excellent aesthetic value depends largely on steel quality. The best flat rolled steels are those that have been decarburized. These steels may be either Open Coil Annealed or Interstitial Free (Vacuum Degassed). Produced specifically for enamel coating and firing, these steels will prevent carbon re-boil (black specking) and fish scaling (glass chipping). The processing and quality of these steels is reviewed.

Vitreous Enameling Steels

Open Coil Anneal (OCA)
Open coil annealing has been used to produce decarburized steels for about half a century. It is unique in that it removes carbon from a solid steel coil rather than the standard techniques of molten steel refining. Done on a relatively low volume basis (coil by coil) it is a very effective process for quality and flexibility of production scheduling.

A standard carbon level heat of slabs is made, then hot rolled, pickle to remove scale then cold rolled to final thickness. Next material is cleaned, then loose wound by inserting separator wires. The coil is then loaded in a special batch-annealing furnace. While coil is annealed, steam is introduced and contacts the steel surface area via the separated coil wraps. Here the steam reacts with carbon in the steel and is taken off in gaseous form. The reaction continues as more carbon diffuses to the surface of the steel. The reaction is deemed complete via dew point or moisture level measurements. The coil is cooled and removed from annealing, then tight wound (separator wire removed). Final carbon level is verified via chemical analysis. Temper rolling imparts coil with desired surface roughness, and material is oiled and sampled for mechanical property testing. This full process flow is illustrated in Figure 1.

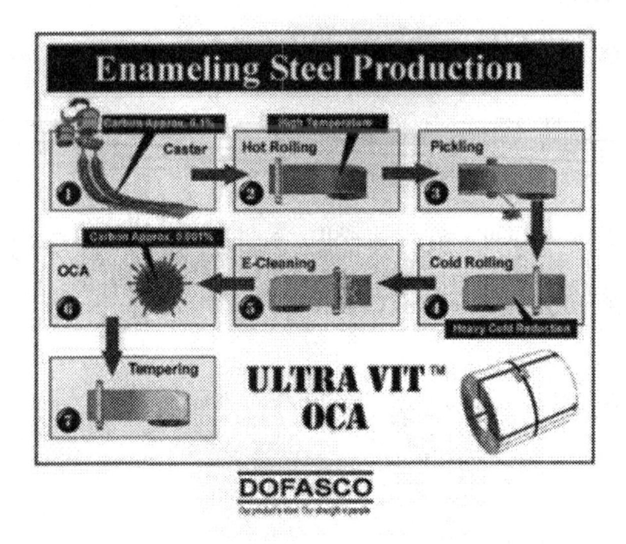

Figure 1 - Enameling Steel Production - UltraVit OCA

Interstitial Free (IF)

Interstitial Free steel was first made during the 1980's. It has found extensive application in the automotive industry due to excellent formability. Its application to decarburized enameling steels is a recent advance as special processing considerations are required. Here the molten steel heat is decarburized by vacuum degassing, followed by stabilizing the residual carbon. Precise chemistry control and uniformity is paramount to successfully make this product on a large volume basis. Once cast, the steel proceeds through regular operations for cold roll steel, hence does not require open coil annealing.

The Interstitial Free steel slabs are hot rolled and coiled at very high temperature. Material is then cold rolled and conventional batch annealed with a long thermal cycle. This produces deep drawing steel suitable for enamel end uses. Typically IF steel has more product size availability than OCA steel, often making it a more suitable choice for very light, or very heavy thickness cold rolled items. Process flow diagram is given in Figure 2.

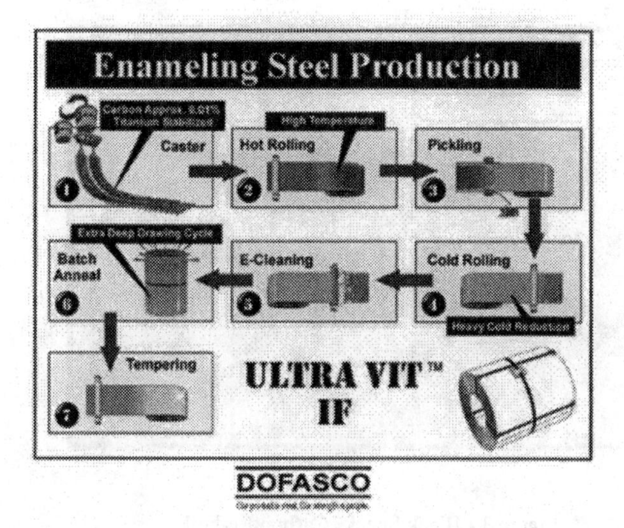

Figure 2 - Enameling Steel Production - UltraVit IF

Quality

Black Speck Prevention

Black specks can appear on enamelware due to numerous sources of contamination. From the steel standpoint, they may result from the firing and breakdown of free carbides in the microstructure. The carbon can volatilize and move up through the molten glass in a turbulent gaseous or volcanic action. This mechanism is illustrated in Figure 3.

For OCA steel the material is decarburized to 0.008% carbon maximum (typically .002%), therefore eliminating iron carbides from the microstructure. Carbon content is confirmed via sheet testing.

For IF steel, the molten steel is vacuum degassed at the ladle metallurgy facility to carbon level of approximately 0.02%. This residual carbon is then stabilized with alloying elements such as titanium. This forms stable titanium carbides in the microstructure that do not break down during enamel firing.

Black speck prevention for the two types of steel is summarized in Figure 4.

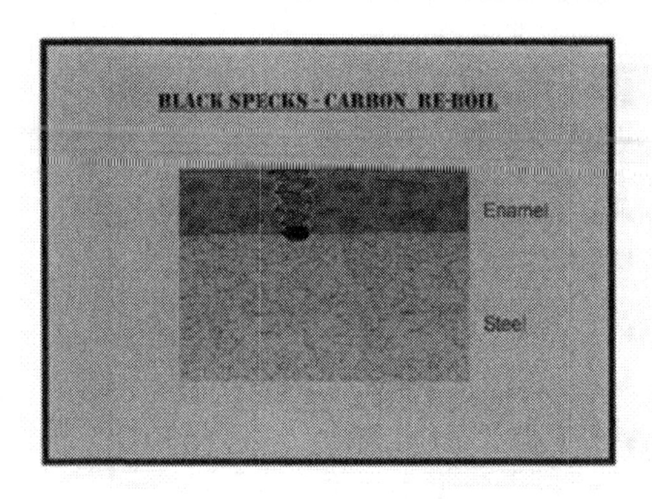

Figure 3 - Black Specks Carbon Re-boil

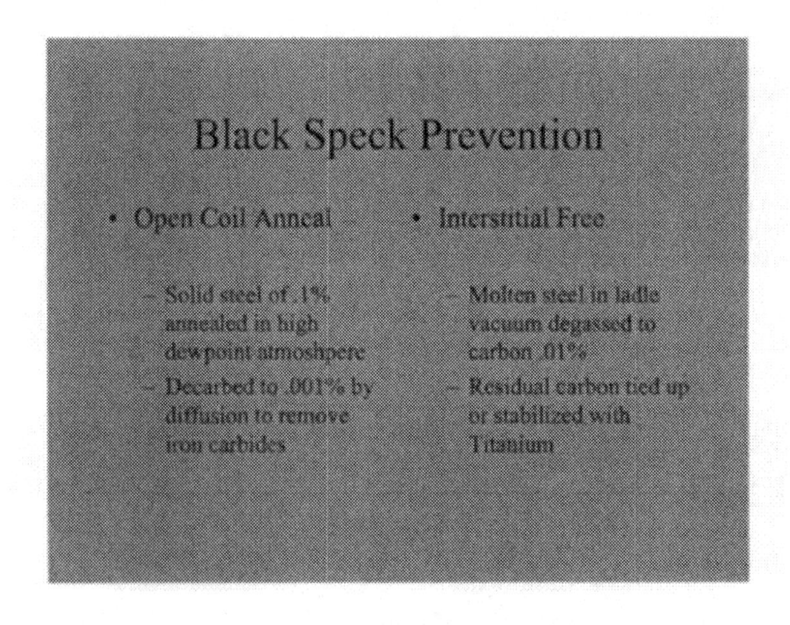

Figure 4 - Black Speck Prevention

Fishscale Prevention
Fish-scale is a very interesting phenomena for steel-enamel systems. During firing, hydrogen gas can enter the steel from the moisture in the enamel frit. The hydrogen is then locked into the steel when the enamel freezes on the surface. Over time this hydrogen tends to migrate to the enamel-steel surface and enough gas pressure can eventually build to crack the glass. This defect mechanism is illustrated in Figure 5. When numerous chips appear on the surface together they reflect light similar to the scales of a fish.

The prevention of fish-scale for both types of steel is predicated on the creation of microscopic voids in the microstructure that can absorb and hold the hydrogen gas indefinitely without detriment to the coating.

For OCA material, the original slab of relatively high carbon is hot rolled and coiled at very high temperature. The resulting large, hard, and irregular shaped iron carbides create steel flow disturbances during subsequent cold rolling. These flow disturbances create the desired voids, and the steel is then decarburized.

For IF material, the slab is again hot rolled and coiled at very high temperature. As such, relatively coarse titanium carbides are precipitated in the hot band. On subsequent cold rolling, numerous voids are created once again by the flow disturbance imparted during heavy cold reduction. A summary of fish-scale prevention is provided in Figure 6.

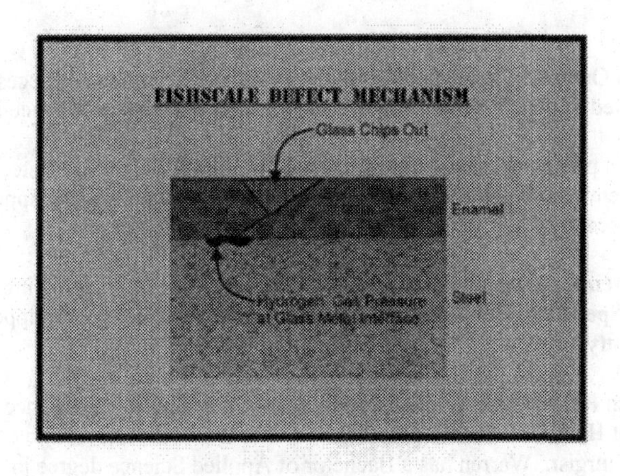

Figure 5 – Fish-scale Defect Mechanism

Fishscale Prevention

Microscopic Voids to Trap Hydrogen Gas

- Open Coil Anneal
  - Iron carbides coiled hot
  - Heavy cold rolling to create voids
  - Iron carbides removed at OCA
- Interstitial Free
  - Titanium carbides coiled hot
  - Heavy cold rolling to create voids
  - TiC remain stable

Figure 6 - Fishscale Prevention

Conclusions

1) Both Open Coil Annealed and Interstitial Free Steels can be successfully applied to glass enameled end uses such as found in the appliance industry.

2) When processed specifically for enameling applications, these steels will prevent both black specking due to carbon re-boil, and glass chipping due to fish-scale.

Acknowledgements: The author wishes to thank John W. Mills - Enameling Steels Manufacturing Specialist, Dofasco, Inc., for his valuable experience and support to the Appliance Industry.

Author: Warren R. Norton, P.Eng. - Technical Service Manager - Appliance for Dofasco, Inc., of Hamilton, Ontario, Canada. Prior to Technical Service he served as Cold Mill Metallurgist. Warren has a Bachelor of Applied Science degree in Metallurgical Engineering from the University of Toronto and is licensed by Professional Engineers Ontario.

A COMPARISON OF ENAMELED AND STAINLESS STEEL SURFACES

David Fedak and Charles Baldwin
Ferro Corporation

Abstract
Several properties important to appliance finishes of porcelain enamel are compared to stainless steel. These include cleanability, scratch resistance, abrasion resistance, heat resistance, and stain resistance. Also, a new stainless-style enamel cover coat is introduced.

Introduction

Stainless steel appliances have become very popular in the last several years because of a consumer preference for a bright finish suggesting sophistication and luxury. Furthermore, high-end products have expanded from a niche at the very top end of the market to the masses.[1] The great majority of kitchen remodeling books seem to promote the use of stainless steel appliances that are ubiquitous on show-room floors.

Unfortunately, the increasing demand for stainless steel in all industries has a double impact on the field of porcelain enameling. First, stainless is purchased by the consumer instead of enameled steel. Second, the high percentage of nickel in the stainless increases costs to the enameling industry. In 2005, about two thirds of all nickel production went into stainless steel and nickel prices were the highest since 1989.[2]

Porcelain (or vitreous) enamel, a glassy coating material suitable for protecting substrates while also decorating them, is a composite material with the best qualities of the metal substrate and the coating.[3] It is generally divided into ground coats and cover coats. Ground coats contain adherence-promoting oxides for good bond to steel and are used in oven cavities, stove grates, hot water tanks, and dual-purpose black enamels. Pyrolytic ground coat is extremely heat resistant enamel for self-cleaning ovens that turn baked-on soils into ash at about 1000°F (427°C). Cover coats provide additional chemical, physical, or cosmetic properties and require a ground coat as a primer layer. Applications include cooktops, washers, sinks, BBQ grills, and architectural panels. Sometimes considered cover coats, aluminum enamels contain low-temperature glasses to fuse directly onto aluminum. These are most commonly used on cookware.

Stainless steel is defined as alloy steel containing more than 10% chromium. A passive chromium oxide layer forms on the surface and provides corrosion resistance. Applications for stainless steel include fasteners, cutlery, flatware, chemical plant equipment, food processing equipment, and kitchen ware.

Stainless steel is classified according to the iron phase present in the alloy. Formed by rapidly quenching austenitic iron from high-temperatures, martensite-containing stainless steels such as alloys 410, 420, or 440C are hard and corrosion resistant. These find use in cutlery, turbine blades, and other high-temperature applications. Ferritic stainless (e.g.,

409, 430, 439, and 444) are more corrosion resistant than the martensitic types and are thus used in nitric acid service, water storage, food processing, automobile trim, furnaces, and turbines. Austenitic stainless steels such as alloys 304, 304L, 308, 310, 316, 316L, 317, 321, and 347 are non-magnetic and offer a good combination of corrosion resistance, high-temperature strength, and ease of fabrication. This type can be strain hardened by cold working. The austenitic 304 alloy is widely used in appliances, and there is also some use of ferritic 430. On appliances, stainless usually has a brushed finish.

Stainless steel is susceptible to the phenomena of sensitization. During exposure between 1000° to 1500°F (538° to 816°C), complex chromium carbides precipitate at grain boundaries. This can lead to unexpected failure through intergranular corrosion. It is eliminated through heat-treating or the use of extra low-carbon steel (e.g., 304L). Weld spots, particularly on ferritic stainless, require heat treatment to prevent this problem.

A number of tests were used to compare the performance of porcelain enamel to stainless steel in several areas that should relate to a consumer's use of a major appliance. Properties investigated were hardness, abrasion resistance, impact resistance, room-temperature stain resistance, and cleanability.

For porcelain enamel, the five enamels shown in Table 1 were run against 304 brushed stainless steel. The four sheet steel systems were all applied dry electrostatically. The 2C1F white cover coat (Ferro PC46C on PL52) and 2C1F biscuit cover coat (Ferro PC168C) are typically used on cooktops. The pyrolytic ground coat (Ferro PL62D) is used on the interior of self-cleaning ovens and has sufficient heat resistance to withstand many cycles of pyrolytic cleaning. The black dual-purpose ground coat (Ferro PL206) is used as a base coat for 2C1F enamels, but it also has a high enough quality surface to be used by itself as black enamel on range exteriors.

Cookware is typically either stainless steel or enameled aluminum. Therefore, aluminum enamel (Ferro GL4317) was evaluated as another alternative to stainless steel.

Enamel	**Substrate**	**Fire**
2C1F white cover coat	Sheet steel	1510°F (821°C)
2C1F biscuit cover coat	Sheet steel	1510°F (821°C)
Pyrolytic ground coat	Sheet steel	1540°F (838°C)
Black dual-purpose ground coat	Sheet steel	1510°F (821°C)
Lead-free aluminum enamel	3003 aluminum	1040°F (560°C)

Table 1. **Enamels tested**

Hardness is defined as a measure of a material's resistance to localized plastic deformation such as a scratch.[4] Different test methods are used for readily scratched materials such as paints, malleable metals, and hard but brittle ceramics. For organic paints, it is evaluated with ASTM D 33630-00 "Standard Test Method for Film Hardness by Pencil Test." For this test, the force required to gouge a coating with a drawing lead of calibrated hardness is assessed.[5] For metals, Knoop, Rockwell or Brinell hardness tests are utilized in which the resistance to deformation with a standard indenter is measured. Ceramics are measured relative to diamond on the Moh's scale. The Knoop, Brinell, Rockwell B, Rockwell C, and Mohs scales are shown side-by-side in Figure 1.

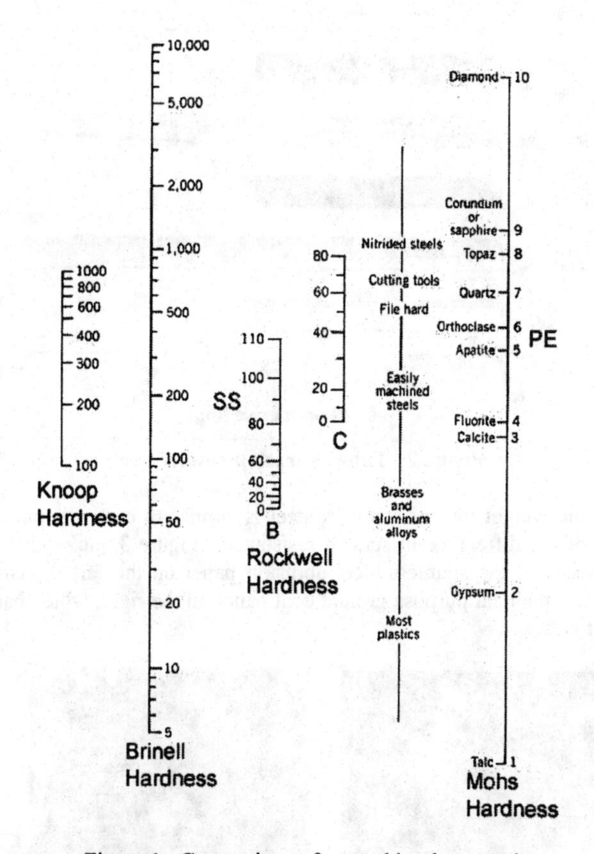

Figure 1. Comparison of several hardness scales

On Figure 1, the point marked SS shows the 88 HRB (Rockwell B) hardness of 304 brushed stainless steel.[6] Point PE shows the hardness of typical porcelain enamel between 5 and 6 on the Mohs scale.[7]

Porcelain was run against stainless on the pencil hardness test. All enamels were off the scale and could not be scratched with the hardest 9H pencil. However, stainless could be scratched with a softer 5H pencil, which correlates with it being softer than enamel on Figure 1.

The abrasion resistance was evaluated using ASTM D 4060-95 "Standard Test Method for Abrasion Resistance of Organic Coatings by the Taber Abraser." Panels were run for 2000 cycles under the hardest CS-17 wheels with a 1 kg load and the weight loss was measured. Data is shown in Figure 2.

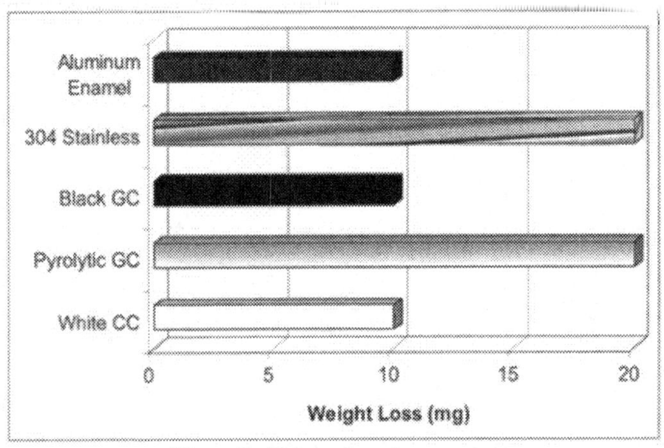

Figure 2. **Taber abrasion results**

Figure 2 shows the weight loss of stainless steel is similar to enamels, but it does not show the effect of the difference in scratch resistance. Figure 3 shows that, due to its poor scratch resistance, the stainless steel abrasion panel on the left is visually much more damaged than the dual purpose ground coat panel on the right, which barely shows any indication of wear.

Figure 3. Panels after abrasion testing

Chipping has been a serious issue for the enameling industry. During impact, stainless steel deforms, and possibly dents but is not seriously compromised. However, the potential for the stainless to be scratched and visually altered permanently is high. For enamels, frit formulations and metal design principles to minimize the chance of chippage are well understood.[8] Furthermore, the porcelain enamel finish readily covers stamping defects and scratches in the metal, which would otherwise be causes of high-cost scrap for stainless steel.

Stain resistance was evaluated for six foodstuffs and three typical household cleaners. The foods were BBQ sauce (Bullseye Original), Heinz ketchup, RealLemon lemon juice, Heinz Worcestershire sauce, Twinings Earl Grey tea, and Heinz vinegar. The cleaners were Easy Off, 409, and (Topco Top Crest) ammonia. Easy Off, containing approximately 5% NaOH, is a very alkaline material with a high pH. Formula 409 contains a mixture of surfactants and a likely smaller amount of NaOH in 2-propanol.

For the foods, soils were placed under watch glasses for 48 hours, while, for the cleaners, the test duration was 24 hours under a watch glass. A value of zero was assigned for no stain and a one was assigned for a permanent stain. The total number of stains was then added up to generate a ranking for the coating.

Results for food-based soils are shown in Figure 4. All of the porcelains except the aluminum enamel were stained by vinegar. The white cover coat and black dual-purpose ground coat were stained with lemon juice, and ketchup also stained the black ground coat. Comparatively, all six soils left light marks on the 304 stainless steel.

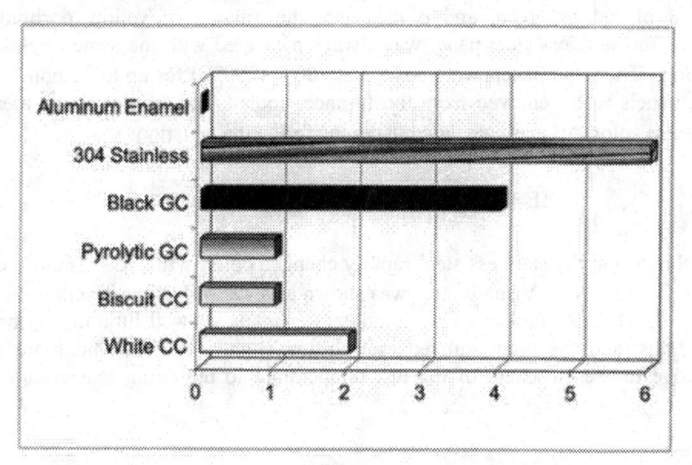

Figure 4. Foodstuff stain results

Results for the cleaner testing are shown in Figure 5. The white cover coat and the aluminum enamel were stained with ammonia. Formula 409 left a permanent mark on the black ground coat. The stainless was attacked by both the Formula 409 and ammonia.

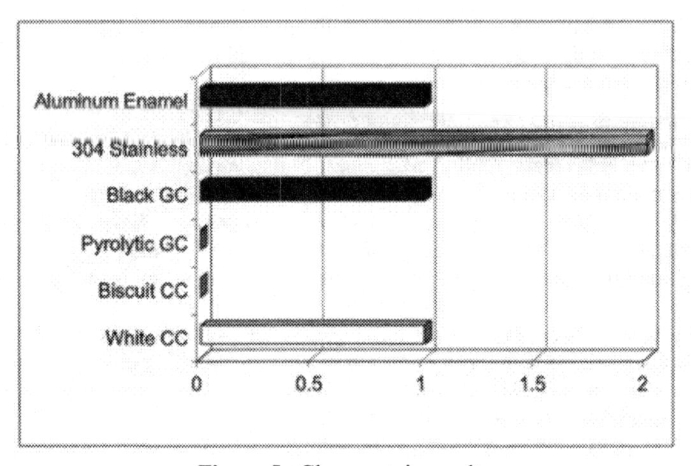

Figure 5. Cleaner stain results

To assess the heat resistance, the Lab color of panels was measured using a DataColor International Spectraflash SF 450 color machine. L measures the brightness, a measures the amount of red or green, and b measures the amount of yellow or blue. For consistency, the stainless steel panel was always measured with the same orientation to the brushing. Then, the panels were soaked at 750°F (399°C) for up to 72 hours. Every 24 hours, panels were removed from the furnace, cooled, and the color was measured. The change in color, ΔE was calculated using the following equation:

$$\Delta E = \sqrt{(L_i - L_f)^2 + (a_i - a_f)^2 + (b_i - b_f)^2}$$

Figure 6 shows that the stainless steel rapidly changed color, with $\Delta E = 16$ after only 24 hours at 750°F (399°C). Visually, this was shown as a sharp shift more yellow (from $b = 4.36$ to $b = 14.75$) and darker. The sheet steel enamels showed little or no change in color readings, and the aluminum enamel shifted slightly brighter and more yellow, probably due to the closeness of the test temperature to the firing temperature of the coating.

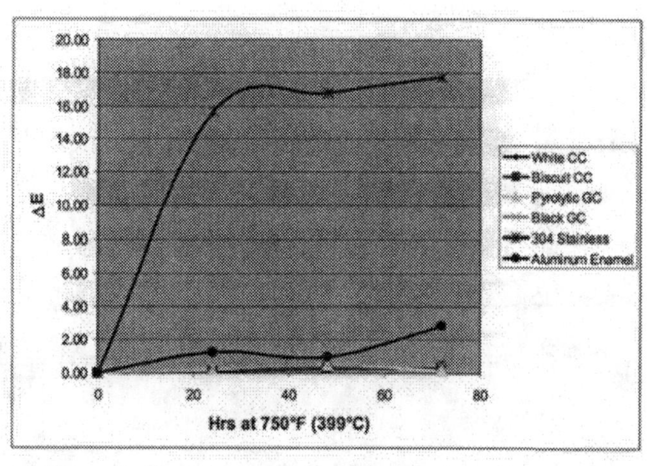

Figure 6. Change in color of stainless compared to enamels

The cleanability was assessed by a modified version of the European FAN (Facile à Nettoyer, literally Easy-to-Clean) test. First, seven steel rings were glued to the coating surface. Second, grape jam, egg beaters, ketchup, salted reconstituted milk, lemon juice, olive oil, and gravy were placed in each ring. Third, the test panel was baked at 450°F (232°C) for one hour. This caused the glue to thermally decompose, allowing the rings to be removed. Finally, the cleanability of each soil was rated using the system shown in Table 2 with a maximum score of 42 for a completely wipe-clean surface and the minimum of 0 for a material to which all of the burned-on residues strongly adhered.

Removal	Stain	Rating
Fell off	No	6
Fell off	Yes	5
Rubbed with paper towel	No	4
Rubbed with paper towel	Yes	3
Steel wool	No	2
Steel wool	Yes	1
Could not remove	Yes	0

Table 2. Cleanability test scoring system

The cleanability results are shown in Figure 7. Grape jam could not be removed from all of the materials. For stainless, the baked-on egg, olive oil, and gravy left stains that were not present on the enamels.

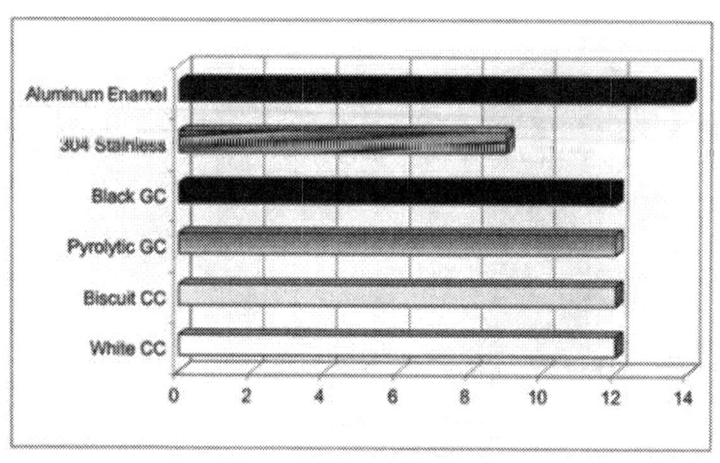

Figure 7. Cleanability results

The comparison testing showed that porcelain enamel is harder to scratch, more stain resistant, much more heat resistant, and easier-to-clean than stainless steel. The results show that stainless steel is functionally inferior to porcelain enamel. Since the consumer preference is for the color, porcelain enamels with the appearance of stainless steel have been developed.

Using metallic pigments, a low-temperature enamel was invented that fires at about 1000°F (538°C) on aluminum or aluminized steel.[9] A high-temperature steel cover coat firing about 1500°F (816°C) has also been developed with similar properties to the cover coats tested. The "stainless-style" enamel is shown in Figure 8 on a range on the backguard, cooktop, and doors.

Figure 8. Stainless-style enamel

In summary, stainless steel appliances have become very popular for a consumer preference for a color that suggests luxury and sophistication. However, the performance of porcelain enamel is superior for major appliances. Enamel is more scratch resistant, more heat resistant, and easier-to-clean. Thus, the stainless-style metallic enamels have been developed to offer the high-end look with the durability of porcelain.

Acknowledgements

Thanks to the PE Development Lab, Harold Pruett, Ken Tyburski, and Brad Devine for their assistance with the high-temperature metallic enamel. Thanks to Renee Pershinsky for proofreading of this paper.

References

[1] Uche Okonkwo, "Brand Papers" *Brandchannel.com* 2005
www.brandchannel.com/papers_review.asp?sp_id=485
(25 April 2005).

[2] "Nickel" *USGS Minerals Information: Nickel*
http://minerals.usgs.gov/minerals/pubs/commodity/nickel/nickemcs05.pdf
(27 April 2005).

[3] Alessando Ubertazzi and Il Plotino, Eds., *Smalto Porcellanato (Vitreous Enamel)*, (Hoepli: Milan, 2002).

[4] William D. Callister, Jr., *Materials Science and Engineering An Introduction*, (Wiley, 1990), pp. 134-9.

[5] Charles Baldwin *et al.*, "A Novel Non-Stick Porcelain Enamel," *Appliance*, 28 – 31 (Nov. 2004).

[6] Russell B. Gunia *et al.*, "Wrought Stainless Steels," in *Metals Handbook Ninth Edition Volume 3: Properties and Selection: Stainless Steels, Tool Materials and Special-Purpose Metals*, edited by William H. Cubberly *et al.*, (American Society for Metals: Metals Park, OH, 1980), p. 19.

[7] A.I. Andrews, *Porcelain Enamels*, (The Garrard Press: Champaign, IL, 1961), p. 546.

[8] William D. Faust *et al.*, "Porcelain Enamels with Improved Chip Resistance," *Proceedings of the 62nd Porcelain Enamel Institute Technical Forum*, 25 – 35 (2000).

[9] Louis Gazo *et al.*, "Porcelain Enamels with a Metallic Appearance," *Proceedings of the 67th Porcelain Enamel Institute Technical Forum*, (In Press), (2005).

CONTINUOUS IMPROVEMENT IN THE PROCESSING OF CERAMIC PIGMENTS

Stan Sulewski
Pemco Corporation, Baltimore, Maryland

Abstract
The process of Continuous Improvement has been used to improve the manufacturing performance of ceramic pigment production. Problems in reproducing the desired pigment color lead to the systematic investigation variable affecting the outcome of production using the Taguchi methodology of experimental design. The fineness of grind, raw materials and temperature of calcination were studied. The results of the work were noted to be closer color to the desired endpoint (master standard), greater stability of the product compared to the master standard, elimination of non-value added steps and identification of better raw materials for processing.

What is Continuous Improvement?

1. Problem solving process
2. Employs up to 20 C.I. Tools

Continuous Improvement Tools

- Brainstorming
- Checksheet
- Flowchart
- Force field

C.A.T. (Corrective Action Team)

a) TS&AD (Technical Service & Applied Development)
b) Weigh and Mix
c) Calcining
d) Quality Assurance
e) Marketing

Problems Identified

1. Difficulty in Reproducing Color
   a) Color too light
   b) Color too blue
2. Raw Material Supply (limited horizon)
3. Non-Value Added Processing
   a) "Sticky" in the Micropulverizer
   b) Wet-milling after calcine
   c) Compromise between L and b
   d) Dried and micropulverized again

Level 4 Taguchi Experiment

- Level 1: Titania A & B
- Level 2: Zircon A & B
- Level 3: Firing Temperature A & B

- #1: Titania Source
- #2: Zircon Source
- #3: Firing

Problems Identified/Solved

1. Difficulty in Reproducing Color
   a) Color too light
   b) Color too blue
2. Raw Material Supply (limited horizon)
3. Non-Value Added Processing
   a) "Sticky" in the Micropulverizer
   b) Wet-milling after calcine
   c) Compromise between L and b
   d) Dried and micropulverized again

Test Procedures

- Wet Milling of Calcine
  - Orbital Mill w/1/4" Media
  - Dried and ground
- Enamel
  - Formula: Frit, Slurry, Oxide: 100 gr. /50 cc /3 gr.
  - Milled 20 minutes w/3/4" Media
  - Wet Application: 1.75 gr. / 2" x 3" GC panels
  - Fired to target temperature and +/- 40 degrees F

Grinding Study

	Virgin	*10*	*20*	*30*	*40*
	8A	8B	8C	8D	8E
L	45.43	44.62	44.05	43.86	44.50
a	0.24	0.37	0.47	0.67	0.70
b	-4.43	-4.51	-4.48	-4.67	-4.36
ΔE		0.82	1.40	1.63	1.04
D(v,0.5)	4.75 μm	3.05 μm	2.02 μm	1.03 μm	0.57 μm
D(v,0.9)	12.18 μm	8.29 μm	6.34 μm	5.14 μm	4.35 μm

C.A.T. Meeting 2

1. Evaluate TS&AD Results
2. Proposed Design Deviation for testing new composition
3. Authorized Auditing of Process
   a) Weigh and Mix
   b) Kiln atmosphere
   c) Kiln profile
   d) Wet Milling
4. TS&AD follow-up tests

TS&AD Tests

1. Fire composition in five profiles
2. Test virgin and various grinds in:
   a) Clear Frit
   b) Semi-Opaque Frit
   c) Opaque Frit
3. Test all of the above to temperature
   a) +40°
   b) -40°
4. Include Master Standard in each system

Conclusions

1. New composition more stable than master standard in each enamel
2. Stability increased with the calcine temperature, particularly in L and a
3. Decreases in calcine temperature had the greatest effect on L
4. Still too blue
5. Wet-milling still required

Effect of Calcine Temperature on Color

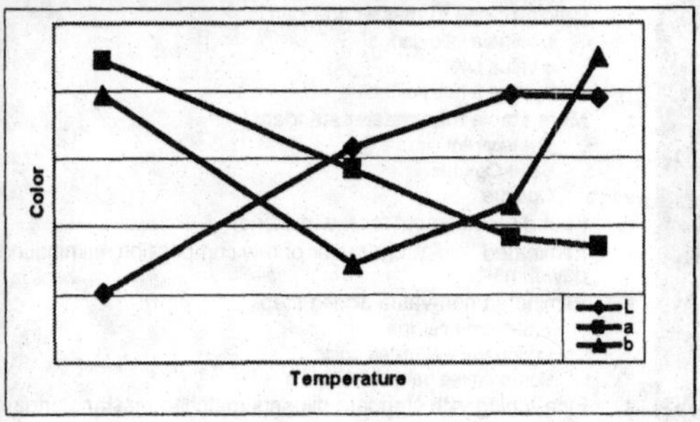

Color
Temperature
L
a
b

TS&AD Tests: Composition Focused

1. Varied existing components
   - With no effect
2. Removed Zircon from calcine
   - Darkened color
3. Added Alumina
   a) Made color yellower
   b) Kept calcine from having a hard sinter

Problems Solved

1. Difficulty in Reproducing Color
   a) Color too light
   b) Color too blue
2. Raw Material Supply (limited horizon)
3. Non-Value Added Processing
   a) "Sticky" in the Micropulverizer
   b) Wet-milling after calcine
   c) Compromise between L and b
   d) Dried and micropulverized again

Summary

1. Calcine close to master standard
   1. L-value a little dark
   2. a-value OK
   3. b-value a half point blue
2. More stable than master standard
   1. Transparent
   2. Semi-Opaque
   3. Opaque
3. Found replacement for raw material
4. Eliminated 'sticky' character of raw composition, eliminating downtime
5. Eliminated non-value added steps
   1. Wet-milling calcine
   2. Drying the wet milled color
   3. Compromise between L and b
6. Fine-tuning with standard diluents matched master standard

HIGH TEMPERATURE CORROSION TESTING OF GLASS LINED STEEL

Gary Griffith
A. O. Smith Corporation, Protective Coatings Division

Abstract
A. O. Smith Corporation produces chemical storage vessels and water heaters. Single and multiple coats of enamels are applied to the metal substrates, typically hot rolled or cold rolled steels. The glass coatings on the steel, porcelain enamels are produced and applied in our facilities. Qualification testing concerning the maximum operating temperature, chemical concentration of solutions in contact with the glass and the subsequent corrosion rate is of great importance.

Corrosion testing methodology has been developed to obtain corrosion rate data for our glasses and then extrapolating this data for service life. A high pressure test cell has been designed which allows exposure of only the enamel coating to the corrosive liquids. Key factors affecting the performance of glass coatings are concentration of the liquid in contact with the glass, the temperature of the liquid, the rate of agitation and any contaminants in the solutions that may accelerate the rate of coating loss.

Why Test?

- Suitability of coating to chemical
- Inspection frequency
- Maintenance requirements
- Determine control point

Test Description

- Maximum operating temperatures
- Chemical concentrations
- Measure weight loss

High Pressure Test Cell

- 1" x 6" Pyrex
- Teflon lined and gasketed

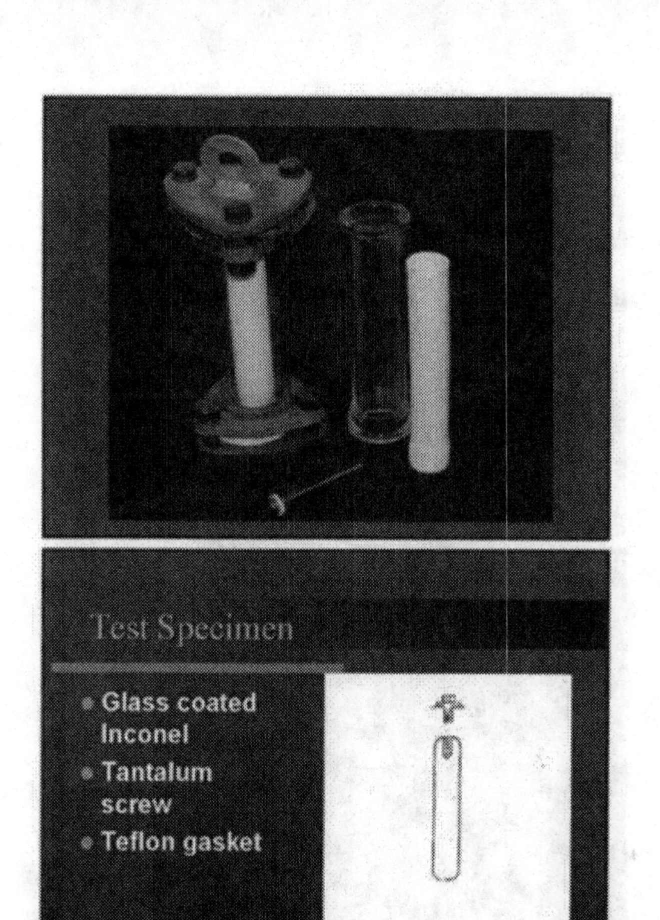
Test Specimen
Glass coated Inconel
Tantalum screw
Teflon gasket

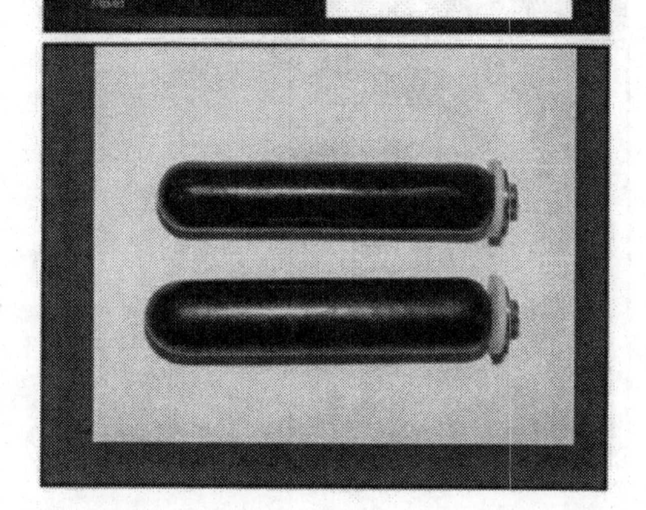

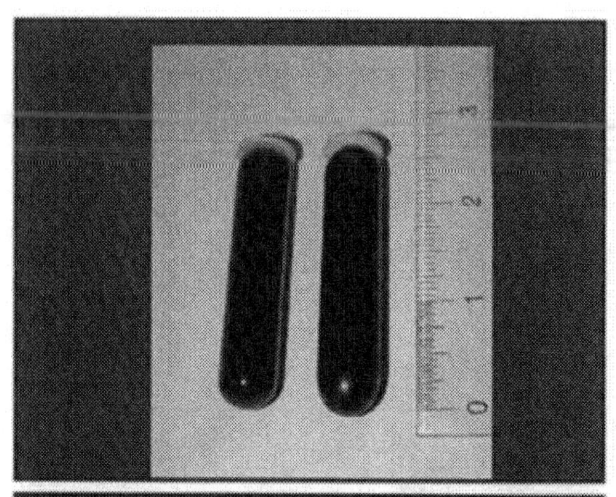

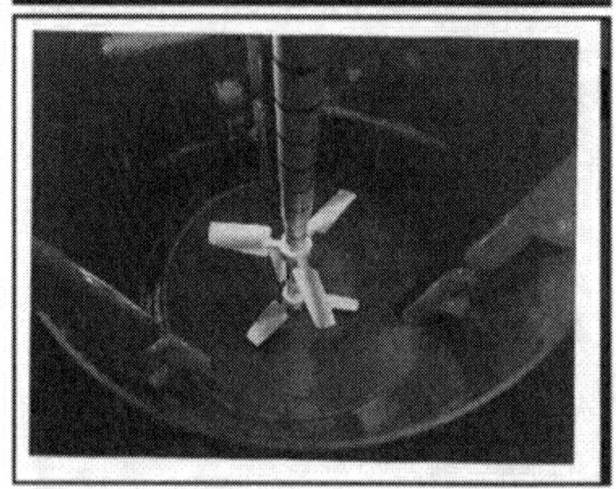

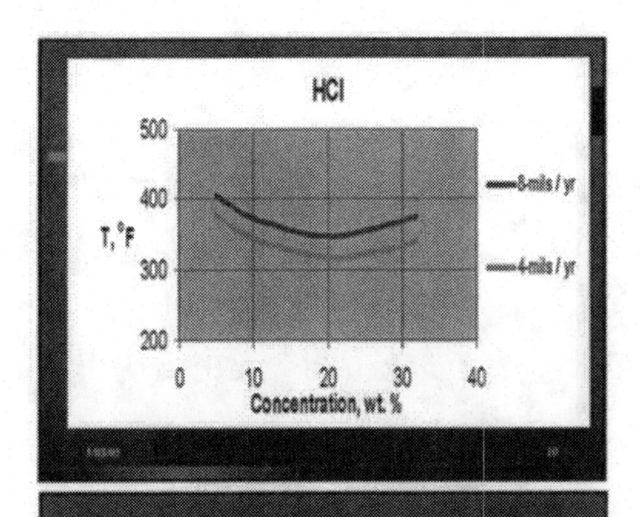

Calculating Mils Lost per Year

- Determine weight loss
- Divide weight loss by area
- Divide weight loss per area by density
- Divide glass lost by time of test
- Convert cm/hr to mils/yr

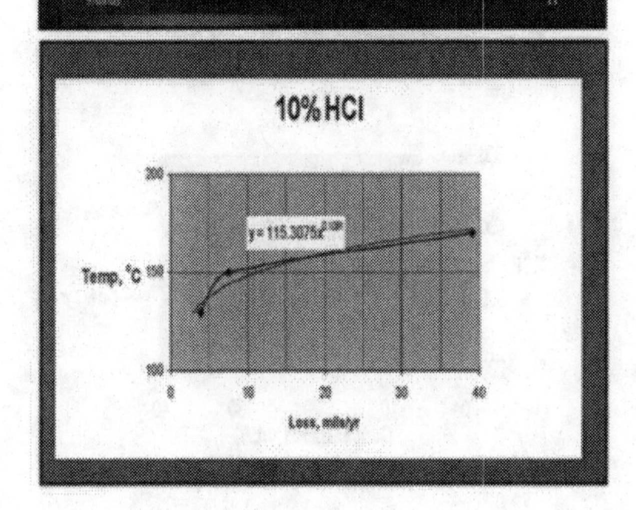

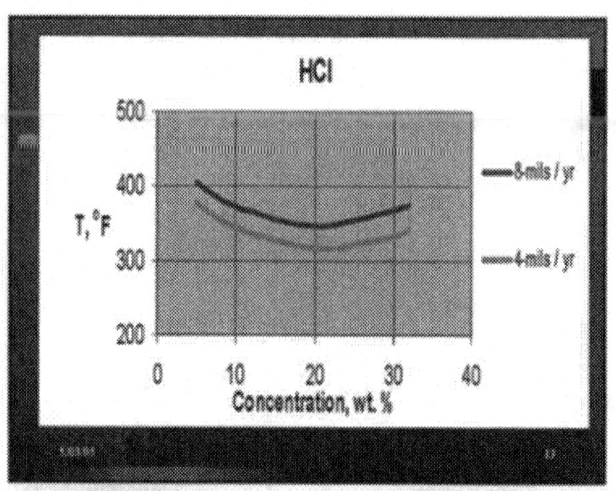
HCl
500
400
300
200
T, °F
0
10
20
30
40
Concentration, wt. %
8-mils / yr
4-mils / yr

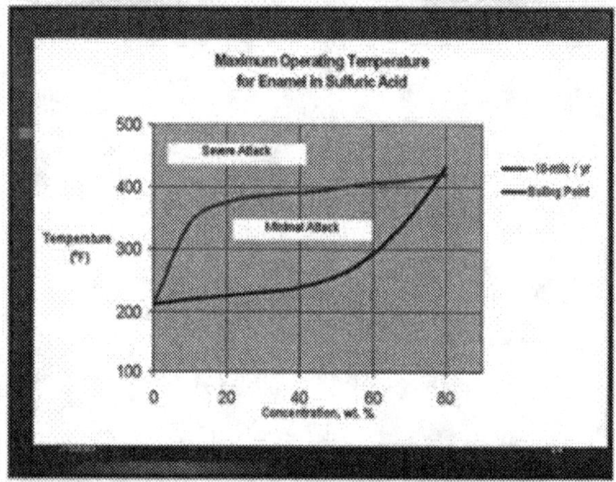
Maximum Operating Temperature
for Enamel in Sulfuric Acid
500
400
300
200
100
Temperature
(°F)
Severe Attack
Minimal Attack
0
20
40
60
80
Concentration, wt. %
Boiling Point

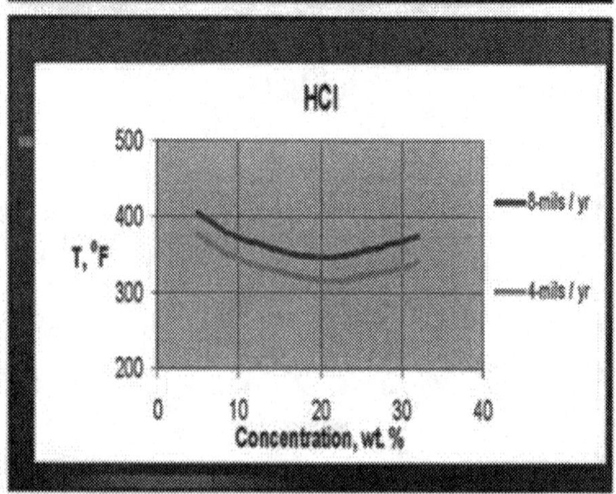
HCl
500
400
300
200
T, °F
0
10
20
30
40
Concentration, wt. %
8-mils / yr
4-mils / yr

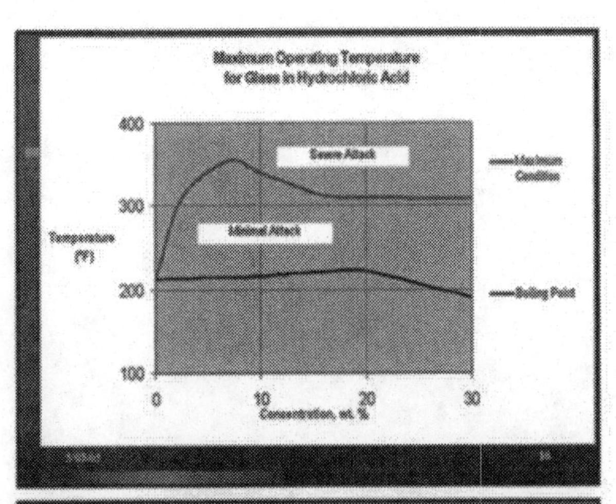

Maximum Operating Temperature
for Glass in Hydrochloric Acid
400
300
200
100
Temperature
(°F)
0
10
20
30
Concentration, wt. %
Boiling Point

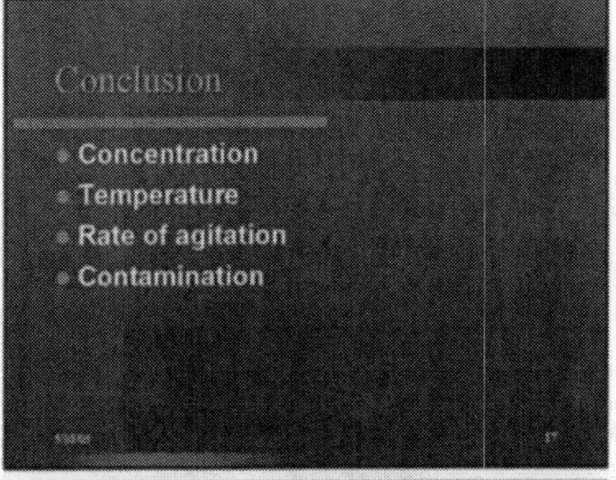

Conclusion
Concentration
Temperature
Rate of agitation
Contamination

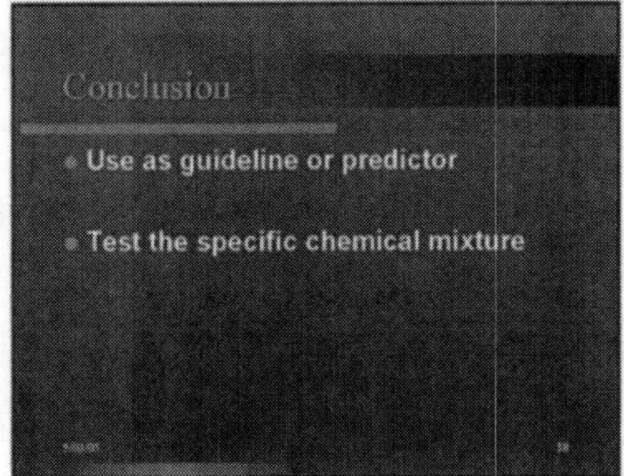

Conclusion
Use as guideline or predictor
Test the specific chemical mixture

POWDER PORCELAIN BOOTH PREVENTIVE MAINTENANCE

Erik Miller
Maytag Cleveland

Abstract

Powder porcelain application has been an industry standard for many years as it provides a high degree of operational efficiency and lowered manpower. In order to maintain the best performance from a powder system, the various components need to be checked frequently. Many application problems can be identified by referring to the Trouble Shooting Guide which outlines problems and sources of deficiencies.

The fluidized bed hopper is the starting point of the powder application process and greatly influences the performance of the application guns. The hopper components such as the level sensors, pick-up tubes, fluidization plate and magnets to catch tramp iron are important items on the checklist. The gun pumps are the heart of the application units and convey the powder from the hopper to the guns. These need to be checked daily for cleanliness and wear. The hoses which are the arteries of the system also need to check daily. Guns need to be checked daily for proper voltage and possible wear as well as weekly tear downs to inspect closely for damage or unusual wear. The booth itself needs to be kept in good order by daily rake downs and complete cleaning monthly, especially the top as dirt and debris could accumulate and fall into the booth during operation. The cartridge filters need to be changed about every six months to insure good air flow in the booth.

During the daily inspections, the Azo screen needs to be inspected for any defects (holes) which might allow coarse particles to get into the powder that is being fed and recirculated into the powder hopper. Additional checks need to be made on a periodic basis of the grounding of the application system (< 1 megaohm), transfer pumps from the booth to the hopper and continuous monitoring of the ambient temperature and humidity. Keeping a stock of needed spare parts will insure that the operation continues with a minimum of disruption.

PEI **ITW Gema**

Porcelain Enamel Powder Delivery Components

and their effect on application

Phil Flasher *ITW* Gema

As many of you know, there are almost countless variables to consider when producing a finished enameled product. Increasing pressure in today's global manufacturing environment requires that we not only work harder but also work smarter. This need for lean manufacturing has resulted in a great number of innovative solutions. A large number of these have can have very attractive payback schedules or return on investment when applied properly. It is my job and desire as an equipment supplier to provide better tools that will make you more successful in your business.

Porcelain Enamel Powder Delivery Components

- The effects of controls and their need in the process
- Powder path components and their effect on application
- These new tools put to work, what it means to the end user

Today I would like to share with some of the more recent advances in powder porcelain application components and their effect on application. I will precede this with an overview of the controls needed to make these tools a successful addition to your application. And finally I will touch on some of the benefits that are experienced with the implementation of these new components.

The effects of controls on the process and their need

- Level controls
- Powder Injector air circuit

I 'm sure those of you currently producing two coat one fire products know that the application process can be quite a cantankerous beast when uncontrolled. Lets first take a look at some areas that play a large role but often get overlooked. Hopper level control is probably top on this list. As simple as it may be, it can have a profound effect on the stability of the powder enamel application.

The effects of controls on the process and their need

- Level controls

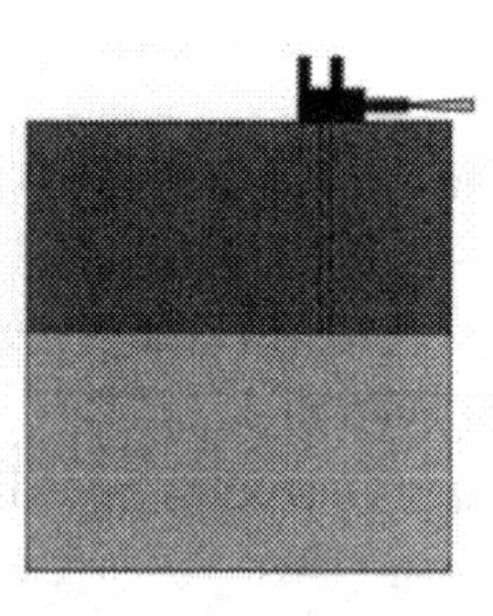

Because the powder is transported by means of a venturi the increasing distance from the injector to the powder as the level rises and falls can have a significant adverse effect on the delivery rate to the gun. Not only does the powder level affect the venturi but it can also cause a fluctuation in material density (powder/air mixture) by influencing the fluidization rate in the hopper.

The effects of controls on the process and their need

- Injector air circuit
- Microprocessor Balanced

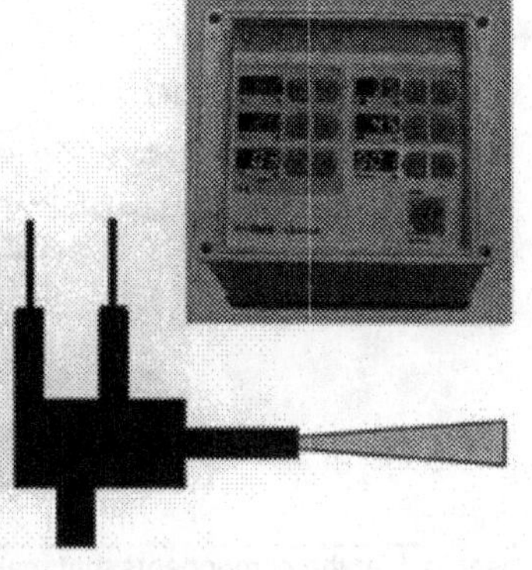

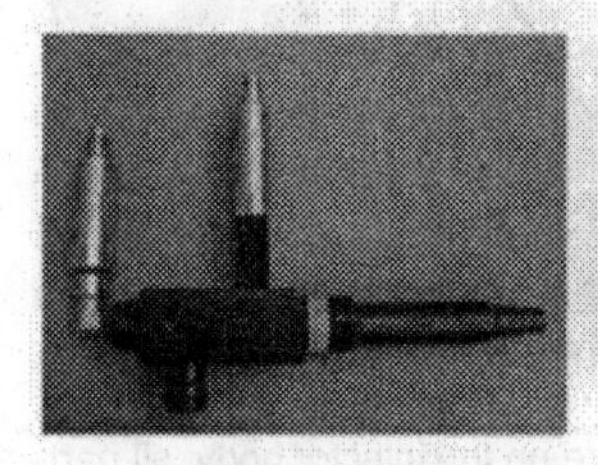

Secondly, the injector and its air circuit make up the heart of process control when it comes to application consistency. The air circuit plays a vital role in the ability to create a stable and repeatable application. This air circuit has evolved from a pressure controlled platform to one of a precisely controlled volumetric delivery. The result comes with many benefits including fewer rejects and rework and the ability to store and recall up to 255 different recipes.

Powder path components and their effect on application

- New injector components and their benefits
- New gun components and charging benefits

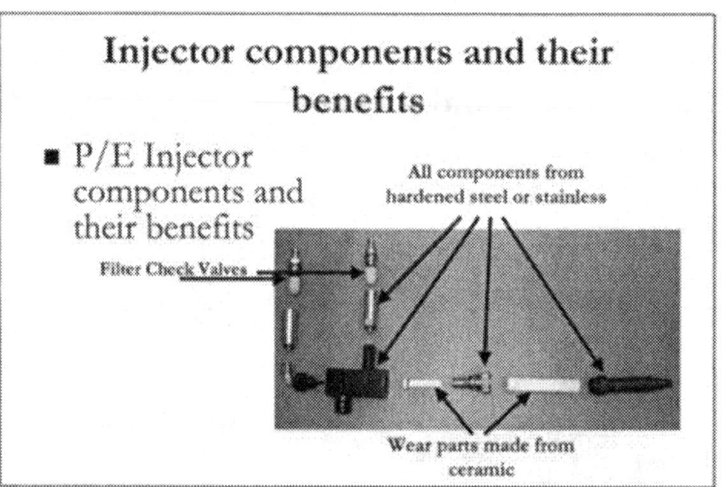

If we take a closer look at the components that make up the injector body, all parts are made from hardened or stainless steel. The wear parts are made from ceramic and provide extended life expectancy. The newly designed check valves are not actually check valves at all. However they serve their purpose even better than a check valve. During spray the dynamics in a powder injector can create an oscillation between the separate air circuits. This can allow frit to migrate backwards toward the precision control unit. The inline filters prevent this while providing low restriction at the higher volumes required to transport porcelain enamel powder.

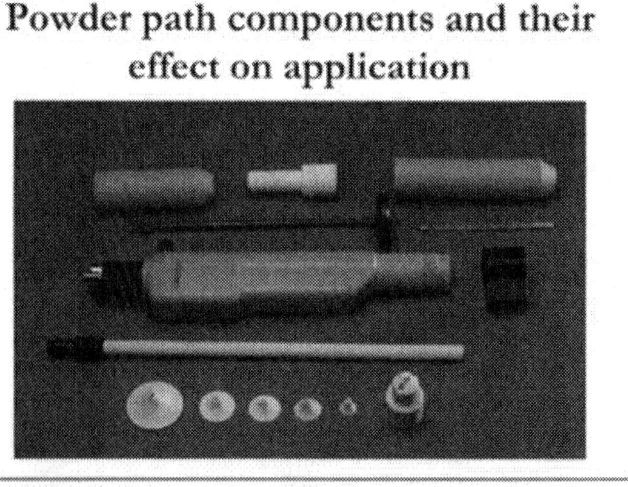

Fundamentally the gun is the same as its predecessor. The components in the powder path have changed the most. A straight through ceramic powder tube is still used but with a larger diameter bore allowing a reduction of velocity in the gun. Unchanged ceramic conical deflectors are used to generate round spray patterns with low velocities to enhance charging.

Powder path components and their effect on application

- New gun components and charging benefits

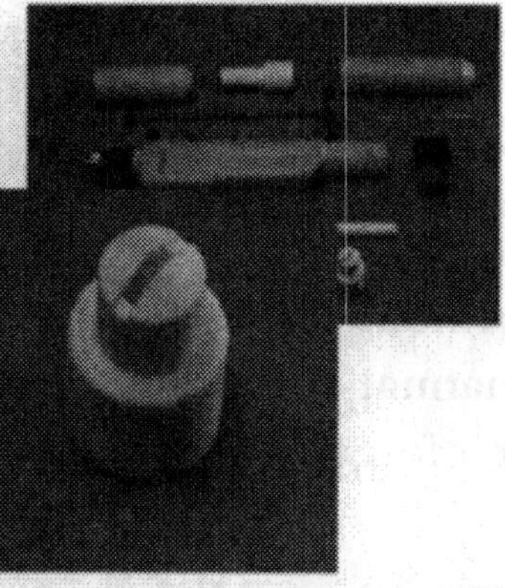

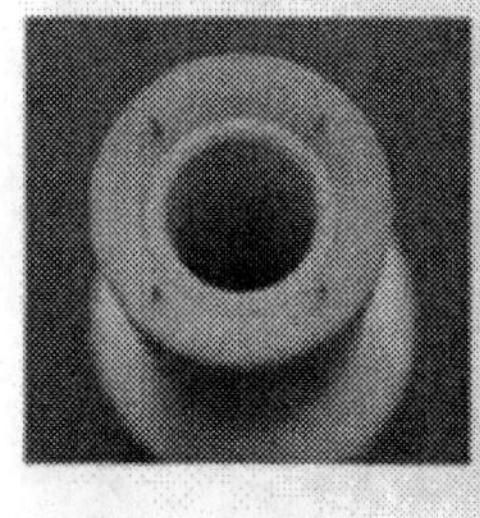

The round spray muzzle continues to use the traditional four electrode design. Typically a multiple electrode design does not have the same charging efficiencies of a single electrode. However most of this is offset by the reduction of powder velocity created by the conical deflector. The additional dwell time in the corona field allows for additional charging by ion bombardment. Flat spray nozzles were much less used in enamel applications because of the inherent design creating high spray velocities due to the high volumes of air needed to transport frit. This is where the new flat spray nozzle has improved.

Powder path components and their effect on application

- New Flat Spray Nozzle

The new nozzle uses an adaptor which expands the internal diameter of the powder path. Anytime a fixed volume of air enters a larger orifice it encounters a reduction of velocity. This reduction coupled with the single electrode charging efficiencies and controls have expanded horizons in automatic powder enamel application systems.

These new tools put to work, what does it mean to the user?

- The ability to coat geometrically complex parts without manual reinforcement

These new tools put to work, what does it mean to the user?

Deep recesses can be coated without excessive enamel accumulation on horizontal surfaces beneath.

These new tools put to work, what does it mean to the user?

- **Process Control**
  - Elimination or reduction of manual reinforcement
  - Consistency and Repeatability
  - Less rework
  - Higher Yields
  - Cost Savings

THE EFFECTS OF DEW POINT ON PORCELAIN ENAMEL POWDER COATING APPLICATION

Frank J. O'Connor, II
Alliance Laundry Systems

Introduction

Electrostatic powder deposition is an efficient process for producing porcelain enamel. But in order to do this while maintaining high quality and low losses requires good control of the correct variables. So this begs the question, "What are the correct variables?" To claim possession of an exhaustive response to this question would be egotistical and false, but one variable that must be included is dew point.

Discussion

Dew point is the combination of air temperature and relative humidity tied together through the Arrhenius equation. Currently air temperature and relative humidity are the most commonly specified and controlled processing parameters. These parameters are typically specified in Cartesian coordinates as Temperature ±°F and RH±%. This can be seen very simply as a "box of control" and it is easy for us to understand. Unfortunately, this is misleading and does not offer a clear picture of what actually needs to be controlled.

To better understand what is happening, let us review the materials in play. We start with a fine powder that is little more than finely ground glass: relatively dense (2.6 g/cc), hard, irregular shape and low dielectric constant (~5). For our discussion we will simply look at the part we are coating as a flat-grounded conductor. None of the powder characteristics assist building or holding powder on a surface. The high density means more mass per layer thickness, which increases the chance of the powder sliding. The hardness and shape means the powder will not pack together neatly; again diminishing the tendency for film build. The final and most damaging characteristic, low dielectric constant, make it disinclined to accept a charge, which is the basic premise for electrostatic powder deposition. With all these factors working against us, why does this process work and what can be done to make it better?

The answer to how we charge our powder is water. Water vapor hydrolyzes on the surface of glass. Because the dielectric constant of water is high (~80) it will take on a charge and thereby hold the charge to the glass. But recognizing that the amount of water on the glass is so small, maintaining a very specific amount of water is critical. If the water content on the glass is too high, powder will build up fast. This is good, but because water is also a conductor the water serves as a grounding path for our charge and thereby neutralizing the charged powder; once the charge dissipates the powder falls off. This is the defect known as sliding. If the powder has too little water it quite simply will not charge. This is known as bluing or light coverage.

So what mechanism do we possess which allows us to so carefully control the amount of water on the surface of the powder we spray? We perform this powder hydrolyses through controlled water vapor density; this is more often referred to as dew point. Dew point can be calculated from relative humidity and temperature, but it is the absolute humidity in the air that corresponds to how much water will attach to our powder. It is helpful to look at the hydrolysis reaction as a probability function; if the particle encounters a water molecule it has a certain chance of bonding up the water. This reaction will be somewhat temperature controlled but much more so it will be controlled by the density of water particles in a given volume.

Relative humidity is the amount of water in the air compared to how much could be dissolved in the air. Unfortunately 'how much can be dissolved', or saturation vapor density, is a function of temperature. That function can best be described by the Arrhenius equation below:

$$K = A \exp(-E/RT)$$

where K is the saturation vapor density, A is a constant, E is an activation energy for water, R is the gas constant and T is the temperature in °K. The important thing to take from this equation is that when the temperature changes by 22°F, the saturation vapor density changes by approximately a factor of 2. So Relative humidity, which we carefully monitor and control, does not offer us adequate control over the hydrolysis of water on the glass.

Now that we see how water on the powder controls adhesion, let's see how the Arrhenius Equation varies the amount of water vapor in our system. The table below shows typical temperature and relative humidity ranges for powder enamel coating operation:

	Low Temp. Limit 69°F	**High Temp. Limit 79°F**
Low R.H. Limit 40%	Saturation Vapor Density=17.8 Vapor Density=7.12	Saturation Vapor Density =24.6 Vapor Density = 9.84
High R.H. Limit 50%	Saturation Vapor Density =17.8 Vapor Density = 8.90	Saturation Vapor Density =24.6 Vapor Density = 12.30

The units on vapor density are grams per cubic meter or parts per million. So this table shows that the total amount of water in the air can vary from 7.12ppm to 12.30ppm. This is a very large change in the water content in the air. This will change the amount of water, which adheres to the powder. And ultimately this will change the coverage properties of the porcelain powder on the enameling iron steel.

Data Review

The data review for this study used actual production loss rates for a three-year period. While the day-to-day results do not show a strong correlation, it can be seen that as the environmental control of the powder coating room came under control, the losses diminished significantly. And in fact it was the better control of the environment that revealed the importance of the dew point measurement. Through historical loss reviews and our better understanding of the importance of dew point, we now have control limits that dictate stopping production based upon the dew point in our powder coating area. We have clearly seen the value of a controlled dew point and the pain of a fluctuating environment. Anecdotal evidence is offered from the shop in the form of color shifts in the base coat powder color. The way the powder adheres to the booth walls can also be correlated to the dew point in the powder room. Unfortunately this adhesion cannot be quantified, but the existence of this phenomenon is clearly visible.

Conclusion

In manufacturing, variation reduction is always the goal. As we struggle to reduce the variation in the powder enamel coating operation, we must determine what measurable factors control how the powder adheres to parts. In three years of looking for the 'big one' it is clear that vapor density, or dew point, is the single largest factor we have been able to identify as impacting our day to day performance. Looking at relative humidity and temperature independently to not explain the variation in the enamel powder coating operation; they must be tied together in order to see the real correlation. The challenge is to stop describing this operation as an art and start defining it as a science. Hopefully others will build upon this and offer further refinement of the enamel powder coating process.

Acknowledgements

The author would like to thank David Conti and David Hickman of Alliance Laundry Systems for their time, effort and ideas in the development of this model. Many thanks go out to Larry Steele (Mapes & Sprowl), Robin Watson (formely Ferro) and Steve Stapleton (Chemetal Oakite) for the many hours of discussion and the many meetings we had working to improve our process. The cooperative spirit of this endeavor made the effort interesting and exciting. I am happy to show our efforts did pay off in the end.

Powder Properties

- Dense (2.6g/cc) – **Requires Large Force to Hold**
- Hard – **Will Not Conform/Pack**
- Irregular Shaped Particles – **Does Not Pack Neatly**
- Low Dielectric Constant (~5) – **Does Not Hold Charge**

Powder Properties Do Not Facilitate Powder Coating Process

So How Does Enamel Powder Coating Work?

- **Saturated Vapor Density** - SVD

The point of maximum moisture in the air for a given temperature

- **Dew Point** – DP

The temperature at which condensation occurs for a given water vapor density

- **Water Vapor Density** - $H2O(g/m^3)$

The mass of water dissolved in a given volume of air

$H2O(g/m^3)$ = SVD X RH where RH is relative humidity

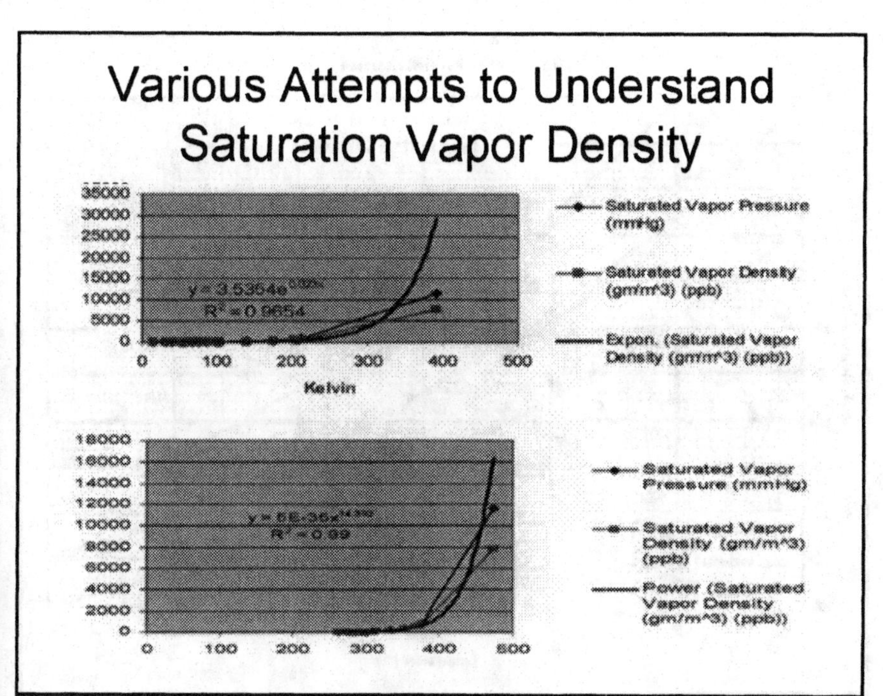

Various Attempts to Understand Saturation Vapor Density
35000
30000
25000
20000
15000
10000
5000
0
0
100
200
300
400
500
Kelvin
Saturated Vapor Pressure (mmHg)
Saturated Vapor Density (gm/m^3) (ppb)
Expon. (Saturated Vapor Density (gm/m^3) (ppb))
18000
16000
14000
12000
10000
8000
6000
4000
2000
0
0
100
200
300
400
500
Saturated Vapor Pressure (mmHg)
Saturated Vapor Density (gm/m^3) (ppb)
Power (Saturated Vapor Density (gm/m^3) (ppb))

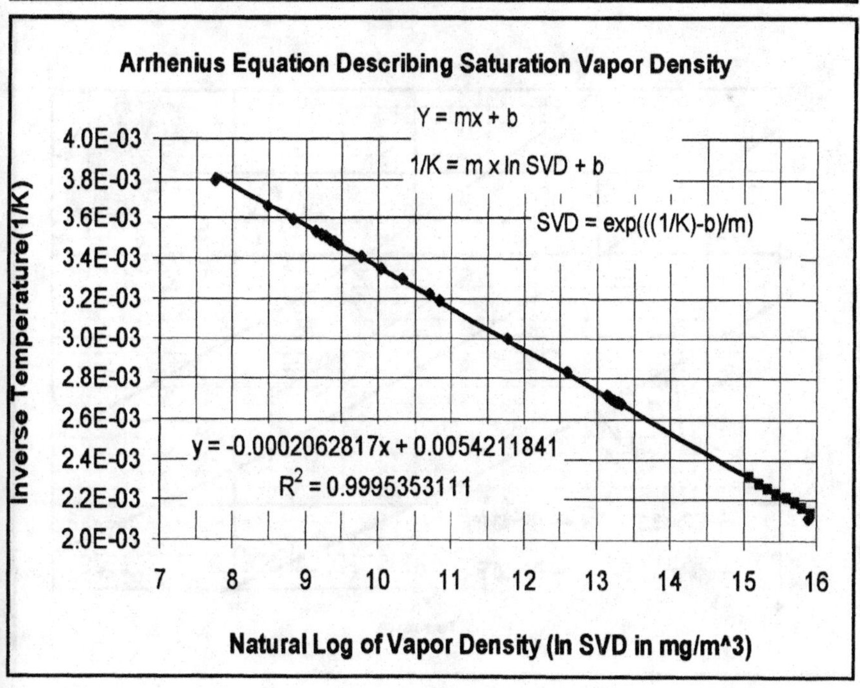

Arrhenius Equation Describing Saturation Vapor Density
Y = mx + b
1/K = m x ln SVD + b
SVD = exp(((1/K)-b)/m)
y = -0.0002062817x + 0.0054211841
R² = 0.9995353111
Inverse Temperature(1/K)
4.0E-03
3.8E-03
3.6E-03
3.4E-03
3.2E-03
3.0E-03
2.8E-03
2.6E-03
2.4E-03
2.2E-03
2.0E-03
7
8
9
10
11
12
13
14
15
16
Natural Log of Vapor Density (ln SVD in mg/m^3)

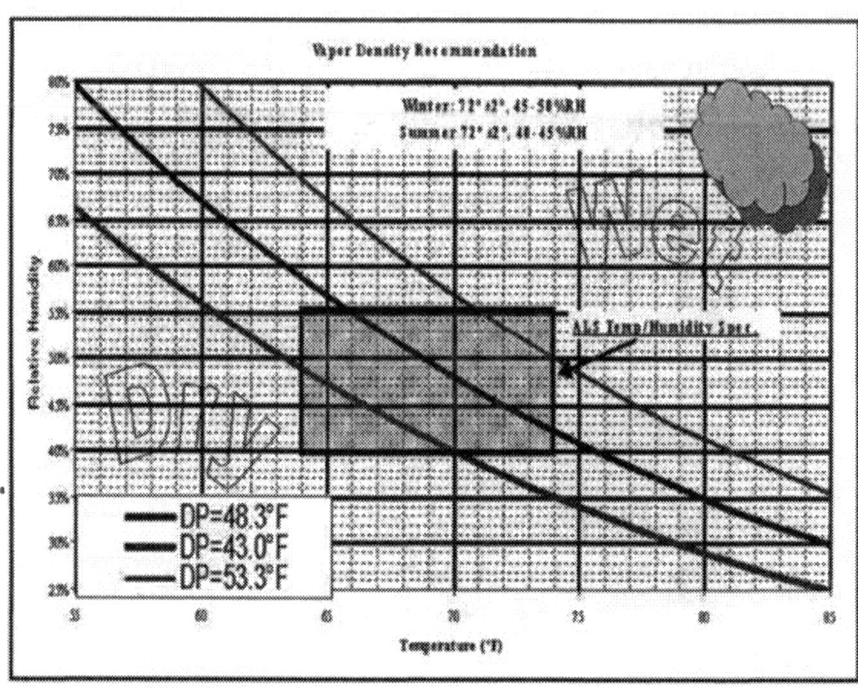
Vapor Density Recommendation
Winter: 72° ±2°, 45-50%RH
Summer 72° ±2°, 40-45%RH
Wet
ALS Temp/Humidity Spec.
Dry
DP=48.3°F
DP=43.0°F
DP=53.3°F
Relative Humidity
Temperature (°F)

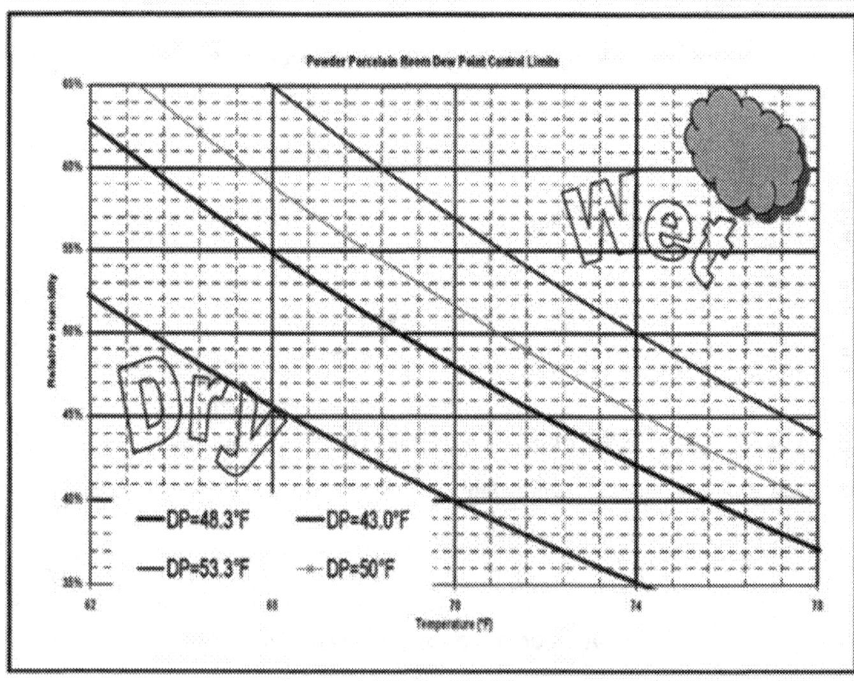
Powder Porcelain Room Dew Point Control Limits
Wet
Dry
DP=48.3°F
DP=43.0°F
DP=53.3°F
DP=50°F
Relative Humidity
Temperature (°F)

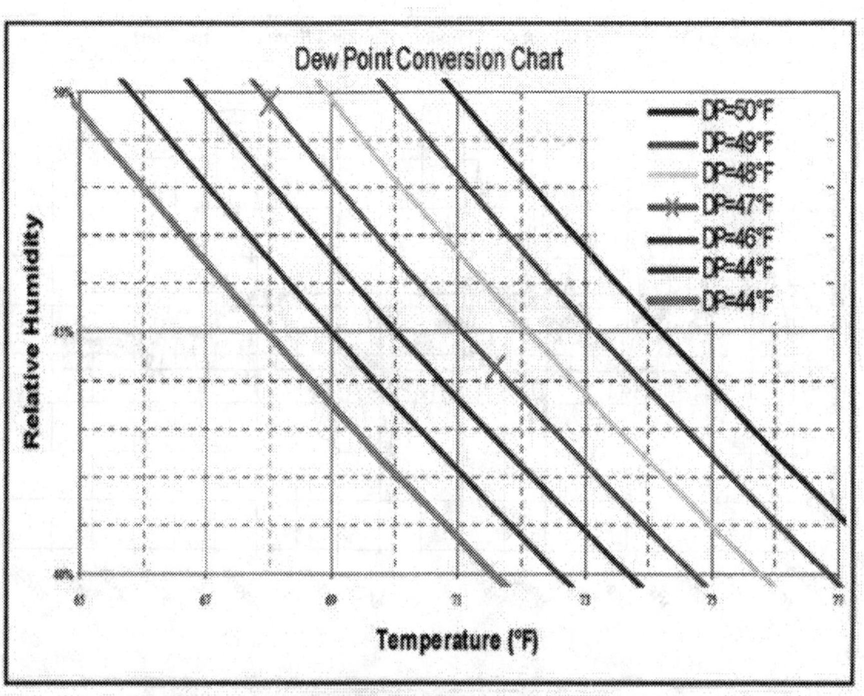
Dew Point Conversion Chart
DP=50°F
DP=49°F
DP=48°F
DP=47°F
DP=46°F
DP=44°F
DP=44°F
Relative Humidity
Temperature (°F)

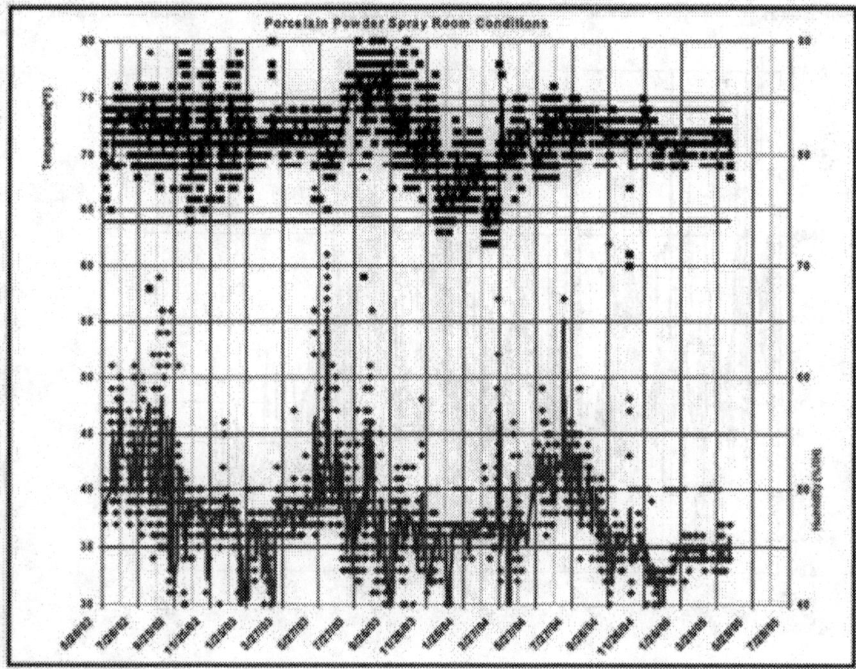
Porcelain Powder Spray Room Conditions

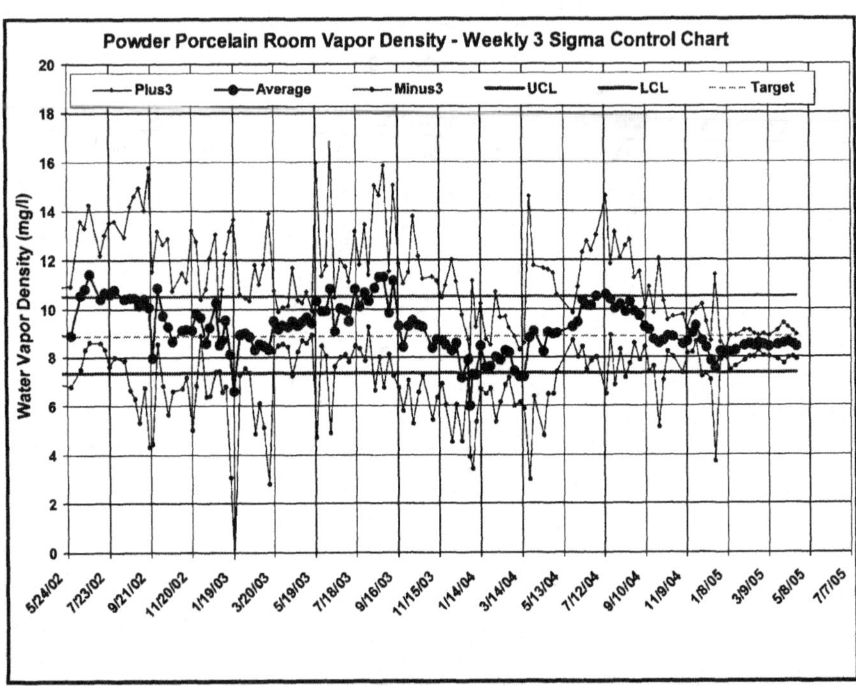
Powder Porcelain Room Vapor Density - Weekly 3 Sigma Control Chart
Plus3
Average
Minus3
UCL
LCL
Target
Water Vapor Density (mg/l)
20
18
16
14
12
10
8
6
4
2
0
5/24/02
7/23/02
9/21/02
11/20/02
1/19/03
3/20/03
5/19/03
7/18/03
9/16/03
11/15/03
1/14/04
3/14/04
5/13/04
7/12/04
9/10/04
11/9/04
1/8/05
3/9/05
5/8/05
7/7/05

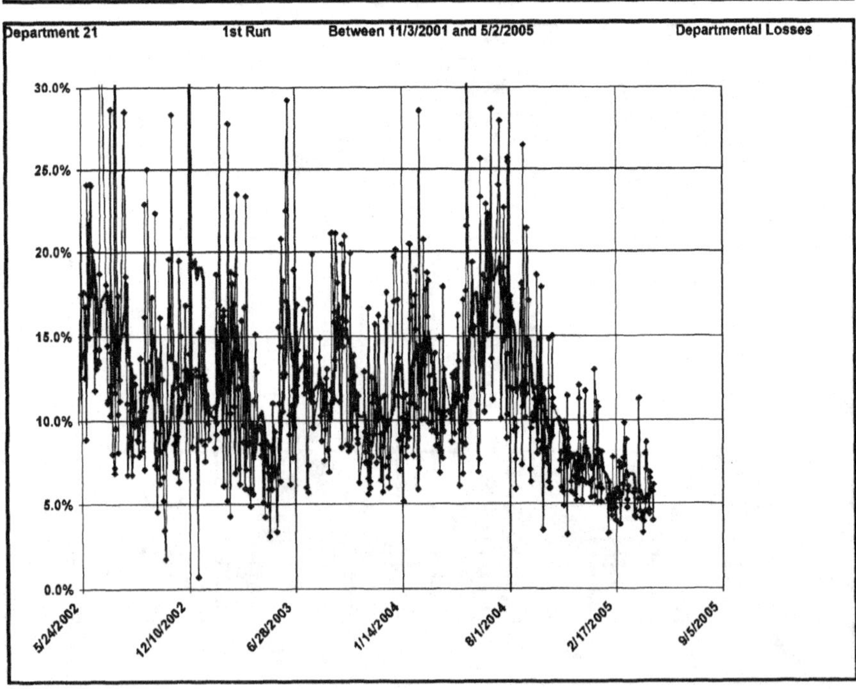
Department 21
1st Run
Between 11/3/2001 and 5/2/2005
Departmental Losses
30.0%
25.0%
20.0%
15.0%
10.0%
5.0%
0.0%
5/24/2002
12/10/2002
6/28/2003
1/14/2004
8/1/2004
2/17/2005
9/5/2005

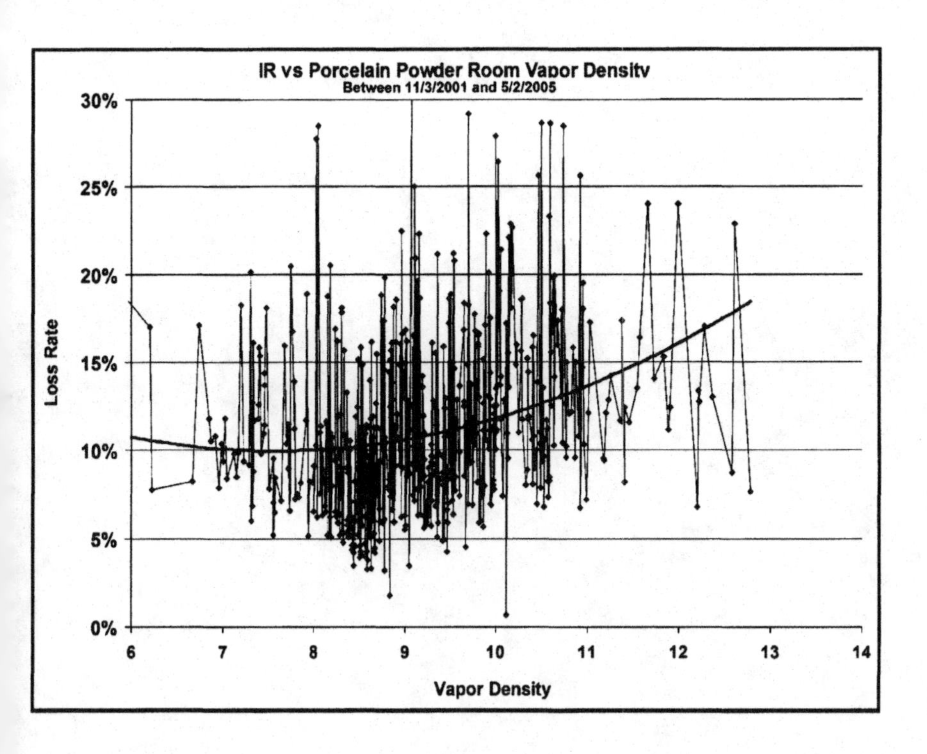
IR vs Porcelain Powder Room Vapor Density
Between 11/3/2001 and 5/2/2005
30%
25%
20%
15%
10%
5%
0%
Loss Rate
6
7
8
9
10
11
12
13
14
Vapor Density

FURNACE RADIANT TUBE COMPARISONS

Ray Gaul
KMI Systems Inc.

Abstract
The heart of a gas-fired porcelain enameling furnace is the heating system, employing radiant tubes in an indirect-fired mode of operation. Over the years subtle changes have been made involving material selection and design configuration. This paper compares and examines these changes, along with some details of operation.

Introduction
Earlier the "Back-to-Basics" workshop explored the various temperature uniformities you can expect with different shaped radiant tubes. Various types and shapes of radiant tubes have been installed in gas-fired porcelain enameling furnaces over the last 40 years or so.

This includes:

1). Vertical Radiant Tubes (VRT) bottom-fired (Wrought)

2). "S" Shaped Tubes (Cast and Wrought)

3). "U" Shaped Tubes (Cast and Wrought)

4). "P" Shaped Tubes (Wrought)

5). Trident Tubes ("W") (Cast and Wrought)

6). Horizontal-Straight Tubes (Cast)

Details of Design
Personally, my first experience with a gas-fired porcelain enameling furnace involved the West Furnace installation at MAYTAG – NEWTON, IA back in 1967. This furnace employed vertical radiant tubes which were bottom fired, approximately 4" in diameter and were positioned on appropriate centers depending on their location within the hot zone chamber.

The hot gases exited out the furnace roof into collector type exhaust manifolds. To increase dwell time and create stratification of the hot gases within the tubes the outside walls of the tubes were "dimpled in" to create eddy currents, which would improve heat release levels. At this point in time, this was the "state-of-the-art" on gas-fired furnaces.

To reduce the number of burners and to improve the "dwell time" effectiveness by changing the configuration, the "S" tube was developed. This particular design provides three (3) legs of radiant heat release, with only one (1) top-fired burner and its shape creates the needed stratification. The "S" tubes empty into the flue gas manifolds located along the floor of the hot zone. The hot flue gases are then directed toward the preheat radiator section. Additional heat release of 20-25 BTU/HR per square inch can be realized from the flue gas manifolds alone. With this particular technology, the ware is preheated providing for an efficient furnace operation.

The "S" tube design is our particular standard at KMI. Initially, cast alloy tubes were installed, utilizing cast "HT" material (35% NI – 15% Chr) with perhaps 3/16" thick walls or greater. This cast version gave way to the wrought or fabricated tube made from #11 ga. (1/8" thick), 330 alloy material, which is the wrought equivalent of Cast "HT". The decision to change to a wrought tube design involved a number of factors. Number one, the wrought version is more ductile, thus the "wish-bone" type of movement that the tubes are subjected to is more readily handled with less distortion and stress. Also the wrought tubes are lighter in weight than their cast counterparts making them easier to handle, and in turn, it takes less energy to bring the tubes up to operating temperature at start-up time. Our experience has shown that "S" tubes fabricated from 330 alloy can have a service life up to seven (7) years with reasonable care and proper burner maintenance and adjustment.

Another shape available is the fabricated "U" tube. It offers a two (2) leg design with one (1) top-fired burner. This configuration can be used in tight quarters, where space is at a premium such as at corners or perhaps in front of an access door. The exhaust leg of a "U" shaped tube can be fitted with a recuperative plug to provide preheated combustion air to the adjacent burner.

The "P" shaped tubes are also fabricated from 330 alloy material once again 1/8" inch in thickness. They are used in conjunction with recuperative-type top-fired burners, and are single-ended radiant tubes transferring waste exhaust heat to the burner air supply at the top end of the tube. In both the "U" shaped tube and the "P" tube system efficiency is achieved by preheating the combustion air.

The trident tube or "W" shaped radiant tube offers three (3) legs of radiant heat release. The two (2) outside legs are equipped with a burner, with the center leg the common exhaust for both adjacent burners. This exhaust leg can be fitted with a recuperative type plug to capture the waste heat and direct it to the adjacent burners. The trident tube can be furnished both in cast versions (HX or HT alloy) or wrought 330 alloy, and can vary somewhat in actual configuration as illustrated by the two (2) versions shown. Again, efficiency is achieved by preheating the combustion air.

Horizontal, straight tubes are usually made from cast alloy such as HT material and are employed on the walls and floors of the hot zone firing from the rear wall of the furnace. They can be as much as 8" O.D. in size and require ladder-type structural alloy supports along their lengths. The tubes themselves must be periodically turned to avoid distortion and sagging.

All radiant tubes have a net release (or transfer rate) of approximately 62.0 BTU/hr per square inch of exposed surface when operating at 1650°F. Typically, the furnace chamber at this point is approximately 1540°F, so we have a temperature head of 110°F during furnace operation.

With any type of combustion system, whether it be "S" tube technology or an alternate design, we suggest that in order to achieve better efficiency that proper loading technique be employed and full-load design levels be maintained. Radiant tubes perform better when pushed a little on demand.

Normally, tube diameter varies from 5 ½" O.D. to perhaps 8" in diameter. Tubes are continuously welded throughout, gas-tight construction. Alloy composition must adhere to ASM and ASTM specifications in all cases, on fabricated tubes. On cast tubes, ACI specifications apply.

In conjunction with the radiant tube service life mentioned above it would be our recommendation to check the tubes at each extended shut down period, concentrating on signs of wear and potential failure. Any severe blistering, warpage / distortion, deformation or cracking should be noted and addressed. During the service life of all alloy components a certain amount of scale is common and metal drop-off is expected, under normal conditions. Radiant tubes begin to show slightly magnetic properties when nickel and chrome have dropped out. When we inspect radiant tubes we use a small magnet to illustrate this change.

Summary

To summarize, the variety of tube shapes that are available for use in porcelain enameling furnaces offer the needed design flexibility to create a uniform heating environment for the ware being fired.

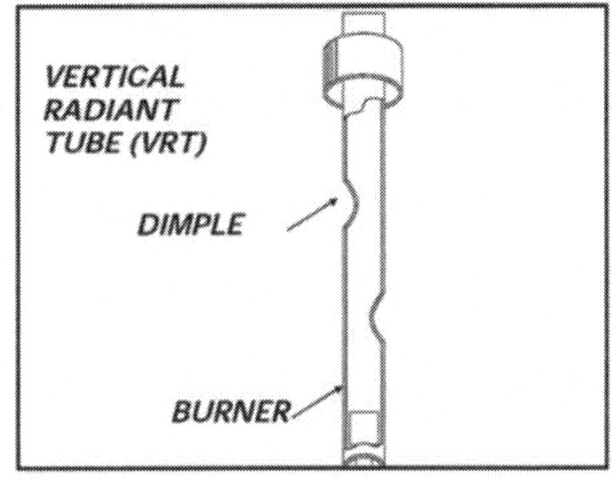
VERTICAL
RADIANT
TUBE (VRT)
DIMPLE
BURNER

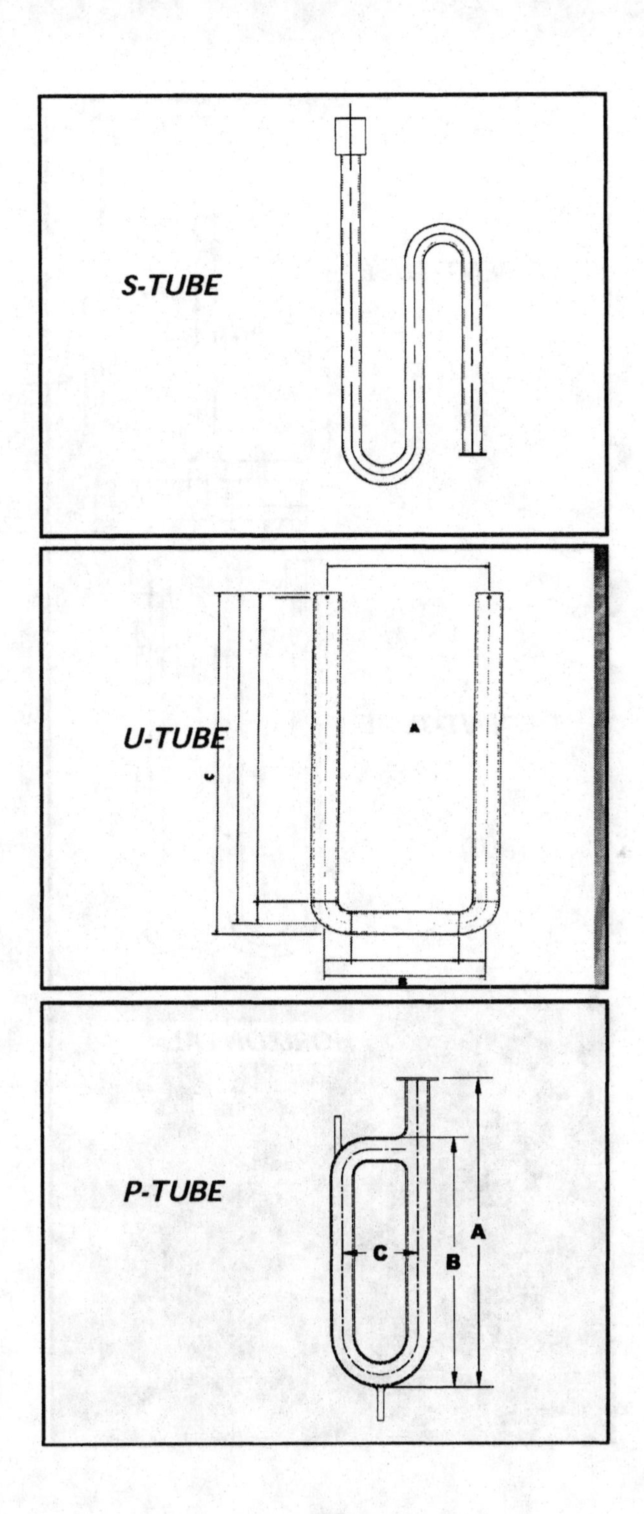
S-TUBE
U-TUBE
A
C
B
P-TUBE
A
B
C

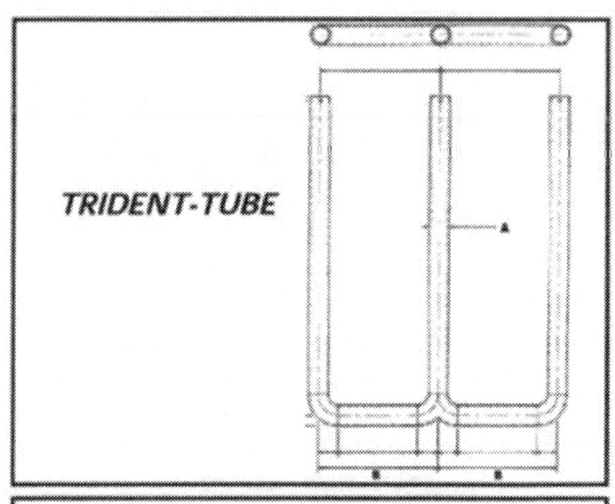
TRIDENT-TUBE
A
B
B

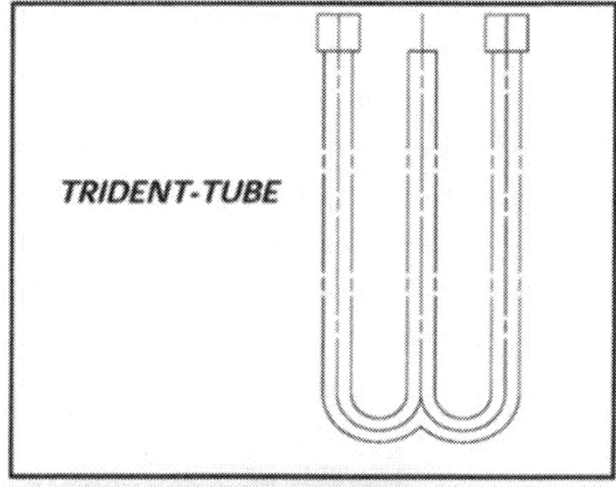
TRIDENT-TUBE

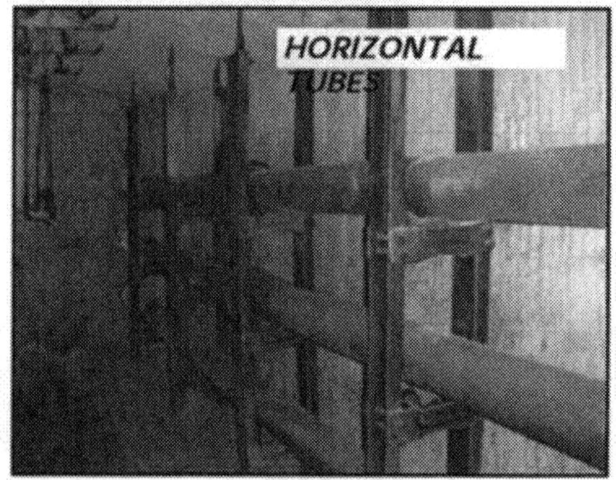
HORIZONTAL
TUBES

EFFECTS OF FURNACE MOISTURE ON ENAMEL QUALITY AND RECOMMENDED LEVELS

Holger Evele
Ferro Corporation

Abstract
The atmosphere of enameling furnaces has a significant impact on the quality of the final enameled product. Means of assessing the moisture levels and how to manage them is discussed.

Introduction

Furnace atmosphere has long been considered a significant contributor to defects commonly seen on porcelain enamel coated ware. A review of previous studies was undertaken and restatement of the recommendations of the Porcelain Enamel Furnace Atmosphere committee follows.

Porcelain enamel chemically bonds to the steel substrate. This process occurs when the iron in the substrate oxidizes at accelerated rates during the fire process.
One of the sources for the oxygen is water present in the coating and in the furnace atmosphere. The following reactions take place:

$$2Fe + 0_2 \longrightarrow 2\,FeO$$

Needed to promote adhesion

$$Fe + H_20 \longrightarrow FeO + 2H$$

Contributes to defects

While taking the oxygen from the water the remaining hydrogen dissolves into the steel. At elevated temperature steel can hold more hydrogen than it can at lower temperatures. During cooling the steel can reach the saturation point for dissolved hydrogen and release hydrogen. Depending on the temperature and viscosity of the glass at the time of release a wide range of defects is possible.

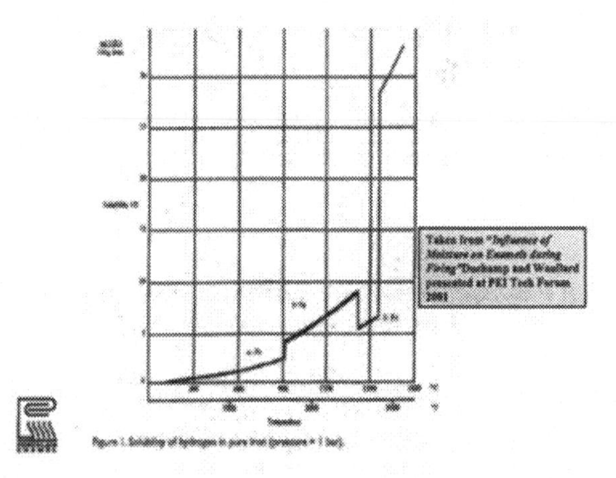

Discussion

Iron must be dissolved into the glass. If no iron dissolves than an oxide layer of weak adhesion is present between the porcelain enamel and the steel. As seen in the micrographs and scans above this enamel has no adhesion and no dissolved iron in the coating.

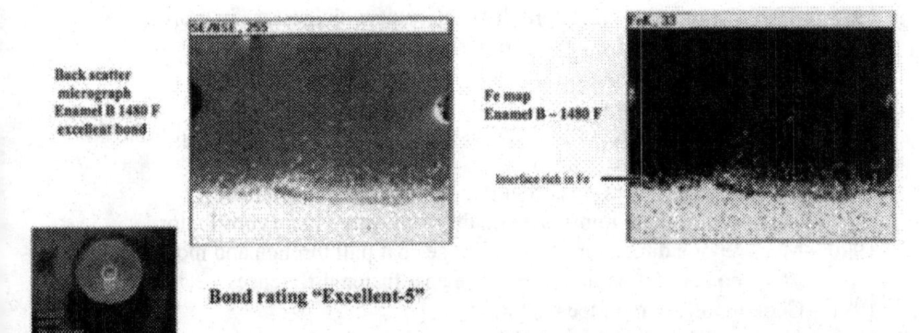

In these micrographs and scans the enamel has excellent fired bond or adhesion and the in the glass clearly shows that iron is dissolving into the enamel coating.

1994 Gordon lists factors that contribute to hydrogen related defects.

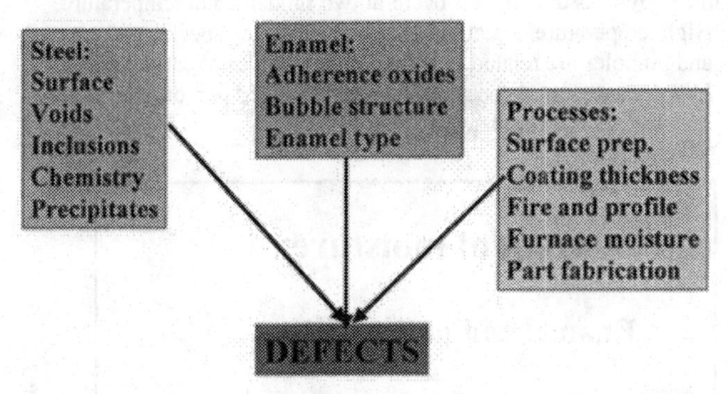

Of all the factors listed by Gordon one that is easy to control and can give wider operating ranges is the furnace atmosphere.

Work from as early as 1938 and as recent as 1993 has shown that gas associated with wide range of defects is primarily hydrogen. While other factors contribute the common denominator is furnace moisture content:

1938 – German investigators Hoff and Klardin extract gas from enameled steel and find it is 85-90% Hydrogen.

1977 – Pensiten reports on impact of moisture to color and surface

Moisture raised by steam injection
from 1.05% to 1.86%
Color Shift in White
L –0.15 a –0.1 b+1.0

1978 & 1992 Ott reports surface problems increase with increase in moisture regardless of enamel composition.

1988 – Steele reports on out-gassing and black specks related to steel composition and moisture content.

Evele/Weinmann report on out-gassing related to cleaning of substrate and and moisture content.

PEI established working committee to investigate on furnace atmosphere impacts on quality – moisture main contributor.

1990 – Moss reports direct correlation of base coat pull through and moisture with no correlation to enamel base coat fusion differences.

1991 – Cook reports committee results:

1 – regardless of enamel used defects eventually seen at high moisture "poor" enamel defects at 1.5% moisture while "good" enamel shows only minor defects at high levels near 7.1% by volume.

2 – The earlier in the fire cycle that peak temperature are achieved the worse the defect levels. Fire cycle can diminish defects but not eliminate them.

3 – Changes in steel composition and substrate preparation impact defect levels.

1993- Joshi – reports that defects occur at two fundamental temperature.
High temperature defects such as blisters, black specks, and copperheads and pinholes are related to hydrogen and oxides of carbon present.
Low temperature defects such as fish scale and pop off are related only to hydrogen content.

Sources of moisture:

Enamel and mill additions

Poorly dried ware

Environment

Furnace POC's

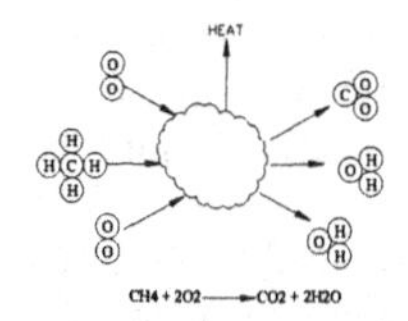

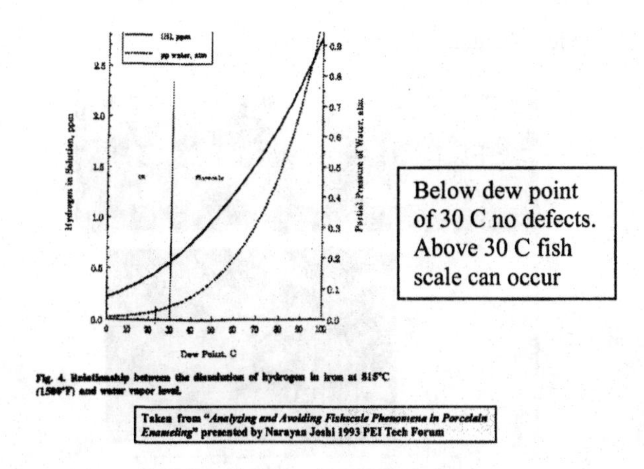

Fig. 4. Relationship between the dissolution of hydrogen in iron at 815°C (1500°F) and water vapor level.

Taken from "*Analyzing and Avoiding Fishscale Phenomena in Porcelain Enameling*" presented by Narayan Joshi 1993 PEI Tech Forum

There are many sources of moisture but perhaps the most critical is the added moisture when by-products of the combustion of natural gas leak into the furnace chamber from poorly maintained radiant tubes or exhaust systems. This coupled with moisture from plant environment can be critical for good enamel quality.

Seasonal Furnace Moisture

Season	Average Plant RH	Average Plant Temp.	Volume Moisture
Spring	60%	70° F	1.4%
Summer	70%	85° F	2.7%
Fall	50%	70° F	1.2%
Winter	30%	65° F	0.6%

Taken from "*Electric Conversion:Two Furnaces-Two Years Experience*" presented by Thomas Pealston 1977 PEI Tech Forum

As shown above contributions from plant environment can but stress on quality even before additional moisture is introduced due to poor maintenance practices.

Possible defects include:

Outgas defect-as glass viscosity doesn't allow gas to escape.

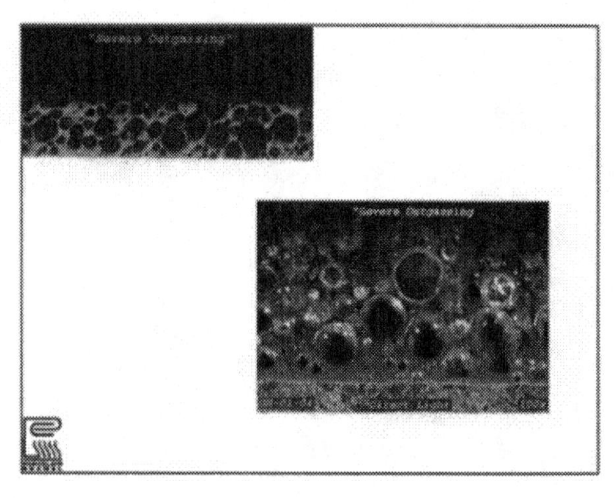

This can be present in a wide range of bubble size and location again depending on glass and fire conditions.

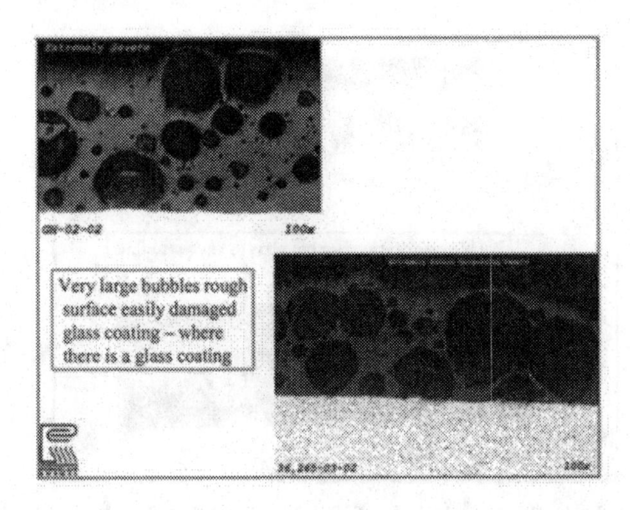

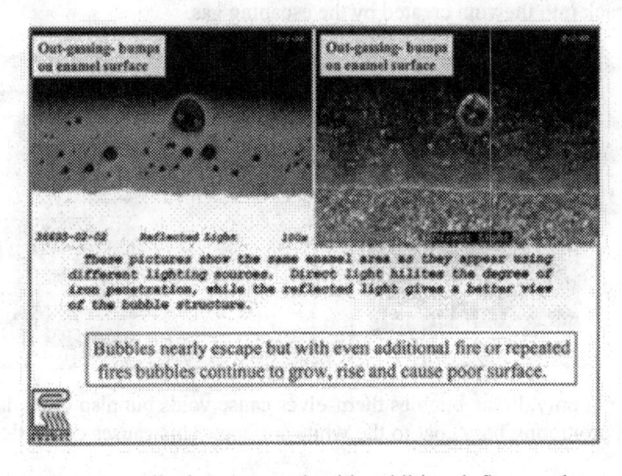

This defect cannot normally be corrected with additional fires and reworking as the bubbles continue to grow and migrate to the enamel surface.

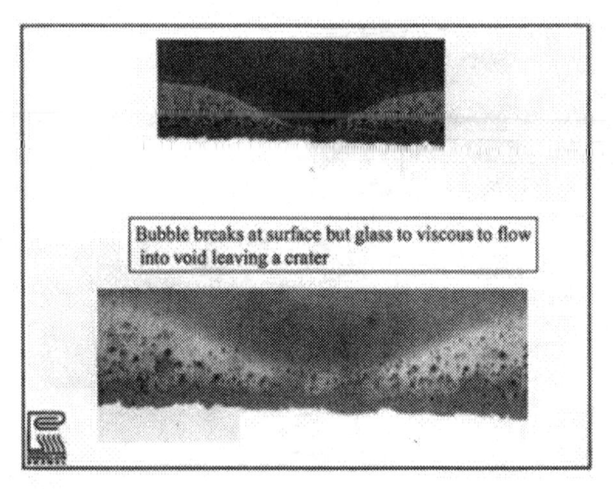

In some cases the bubbles reach the surface and rupture but the glass viscosity, in this case due to the crystallization of the titanium white cover coat is too high to allow the glass to flow back into the void created by the escaping gas.

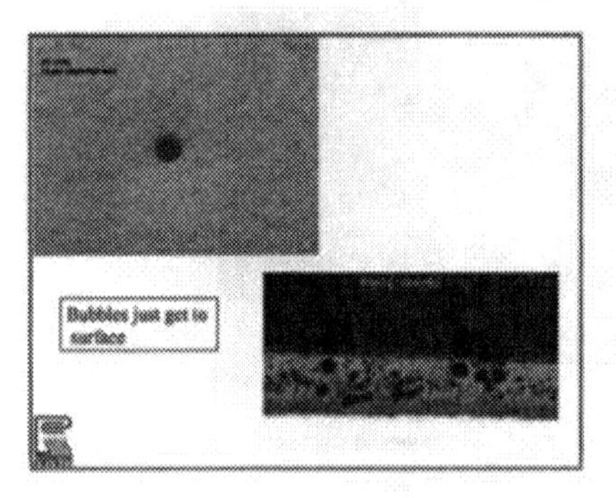

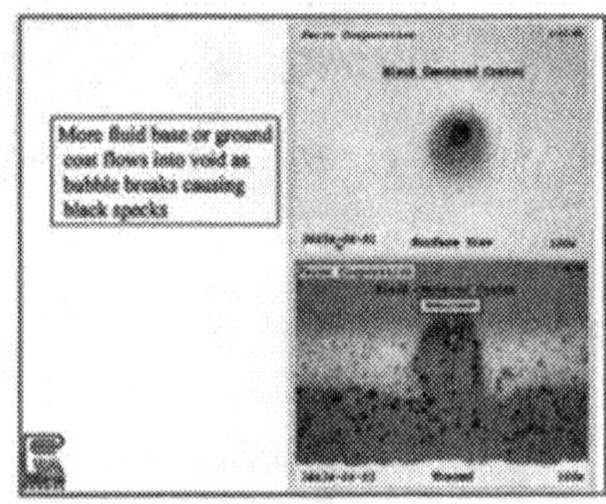

In these cases not only do the bubbles themselves cause voids but also draw dark colored iron rich bond promoting base coat to the white surface. This causes cosmetic rejects for black specks.

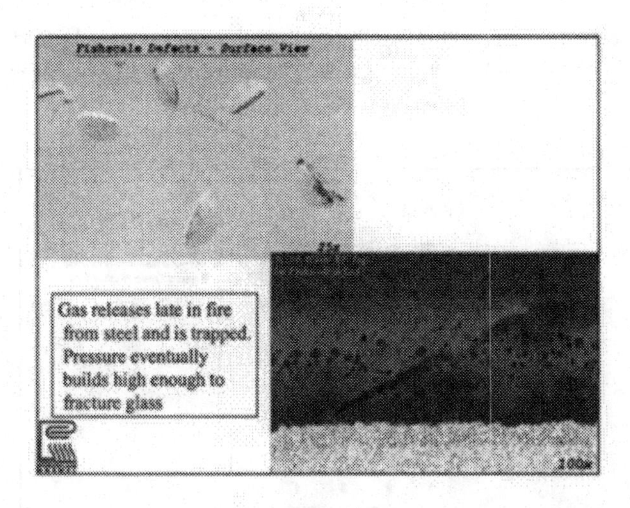

One of the worst defects is fish scale. Gas remains trapped near the glass/steel interface. Once sufficient pressure builds the glass fractures forming small chips that look like fish scale. This is both a cosmetic and durability defect as the coating is no longer continuous.

Conclusions

The Porcelain Enamel Committee on Furnace Atmosphere conducted a wide range of trials with the following results.

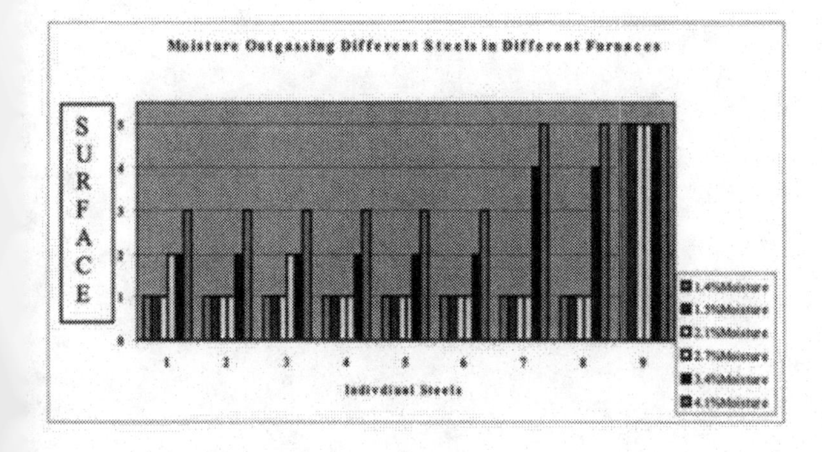

A range of steels, cleaning, and different fire conditions where tested. Surface quality was given an assigned value. Once these numbers were compared to moisture content the following correlation analysis was performed.

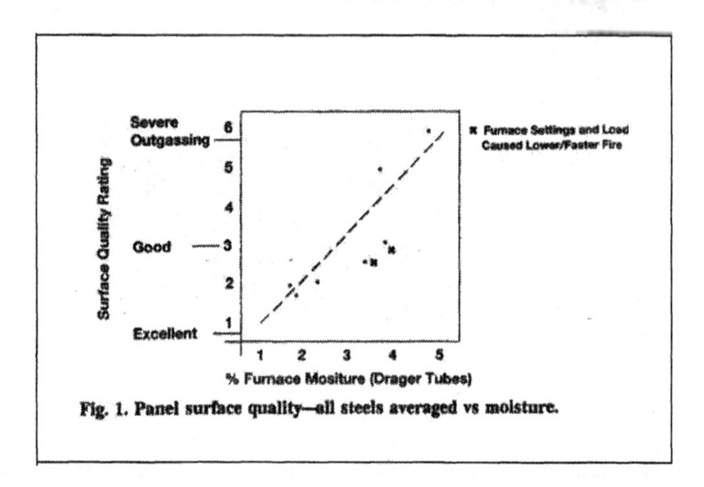

Fig. 1. Panel surface quality—all steels averaged vs moisture.

From this graph and all previous efforts it is clear that maintaining a reasonable moisture level is critical when trying to achieve high quality yields during porcelain enamel coating of ware. In conclusion - while success is still somewhat dependent on factors beyond moisture that the most robust processes tend to have moisture content below 2.5% volume percent of moisture.

CONTINUOUS MEASUREMENT OF FURNACE MOISTURE

Peter Vodak
Engineered Storage Products Company, DeKalb, IL

Introduction

Furnace moisture level has been shown to be critical for the proper firing of porcelain enamel coatings[1,2,3]. The PEI Technical Manual states that the proper moisture level is between 1% and 3%, with the ideal level being approximately 2%[4]. While one may know what the desired level is, measuring and controlling the furnace moisture level in production has been a difficult proposition.

This paper will cover the monitoring of furnace moisture level and the control of furnace moisture level on a continuous basis. The portion on monitoring will begin with a discussion of moisture measurement theory and then move into practical measurement techniques.

Furnace Moisture Measurement – Theory

To discuss moisture measurement it is first necessary to review a few key concepts.

Relative Humidity

Relative humidity is commonly used to measure moisture level for many applications in everyday life. It is defined as follows:

$$RH = P_W/P_S \times 100\%$$

Where

RH = percent relative humidity
P_W = partial pressure of water vapor in the air
P_S = saturation pressure

Relative humidity does not work well for moisture level measurement above 212°F. At or below 212°F the maximum achievable relative humidity is 100%. Above 212°F, at 100% water vapor P_W is equal to atmospheric pressure, while P_S rapidly increases above atmospheric as the temperature increases (because water boils at 212°F). Therefore, above 212°F at atmospheric pressure the maximum relative humidity is less than 100% even if the air is composed of 100% water vapor. This is demonstrated in Table 1.

Table 1 – Maximum Achievable RH as a Function of Temperature

Temperature	Maximum RH Achievable At Atmospheric Pressure
212°F	100%
300°F	20%
400°F	5.9%
700°F	0.48%

One can see that relative humidity is not a good indicator of moisture level for the high temperatures used in firing porcelain enamel. Therefore, we need to use a different approach.

Percent Moisture by Volume

Percent moisture by volume (MV) is an absolute moisture scale. It is not affected by changes in temperature the way that relative humidity is. It is defined as follows:

$$MV = N_W/N_T \times 100\%$$

Where

MV = percent moisture by volume
N_W = number of water molecules per unit volume
N_T = total number of molecules per unit volume

For example, an MV of 50% corresponds to 500,000 water molecules in 1,000,000 total molecules.

Dew Point

The dew point is the temperature where the relative humidity reaches 100%. It is also the temperature at which condensation occurs. There is a correlation between the dew point and the moisture level in the air.

Current Moisture Measurement Techniques

Dräger Tube

In the Dräger tube method of moisture measurement, a known volume of air is drawn across a known amount of a hygroscopic chemical. The chemical is stored in a sealed glass tube, and the ends are broken off just prior to taking the measurement. The moisture in the air reacts with the chemical, which then changes color. The amount of the chemical which changes color is measured and indicates the amount of moisture present in the air.

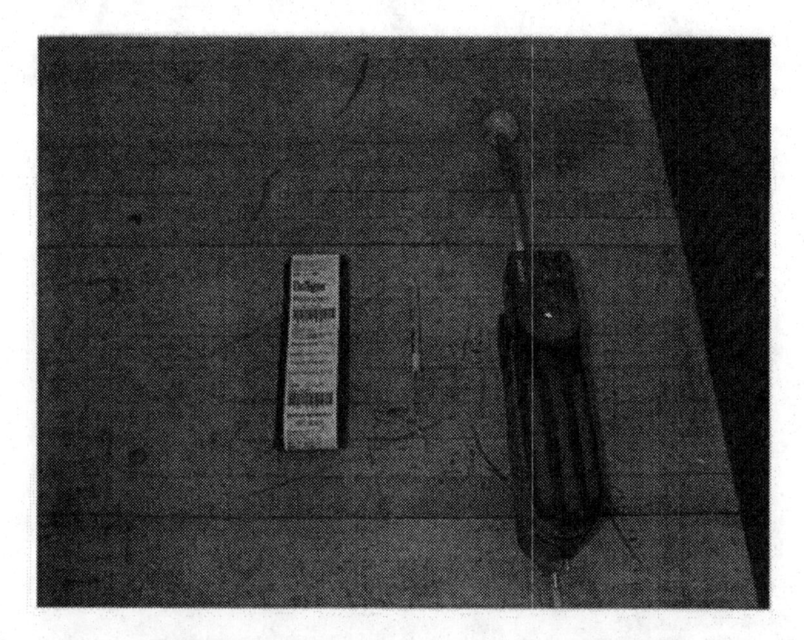

Figure 1 – Dräger Tube With Pump and Stainless Steel Extension

The Dräger tube method gives percent moisture by volume (MV). It is run manually by an operator and requires careful technique to achieve consistent results. The method gives a rapid response (within 30 seconds). The Dräger tubes are relatively expensive (approximately $10 apiece). In the case of furnace moisture measurement the air sample is drawn directly from the furnace from a sample port using a stainless steel extension probe that connects to the drawing pump.

Condensing Mirror

In the condensing mirror method, a known volume of air is drawn into a sealed chamber containing a mirror. The pressure is raised to a known value, and the air is suddenly expanded by means of opening a valve. The operator looks through a window to see a cone of fog develop (condensation occurring). The pressure is then varied and the test re-run. Multiple readings are required to determine the endpoint where fog occurs or does not occur.

Figure 2 – Condensing Mirror Unit With Draw Tube

The condensing mirror method gives dew point, which as has been previously stated correlates to percent moisture by volume. It is performed manually by an operator and it too requires careful technique to achieve consistent results. The method takes a bit of time to make a determination (approximately 5 minutes with a trained operator).

Continuous Moisture Measurement

While both of the methods listed above do work, both require skilled operators to manually run the test. The Dräger tube also has an on-going cost in replacement tubes. Additionally, neither of these methods provides continuous moisture level measurement. Engineered Storage Products Company (ESPC) therefore sought to come up with a method for continuous furnace moisture level measurement that would meet the following criteria:

1) Be compatible with the furnace environment at 1600°F (continuous).
2) Monitor furnace moisture level on a continuous basis without direct operator involvement.
3) Give repeatable results that correlate with the other two methods.

A high-temperature continuous moisture analyzer[a] was tested. It is a solid-state device that utilizes a variable capacitance sensor to determine percent moisture by volume.

[a] MAC 120 moisture analyzer made by MAC Instruments, Sandusky, OH. Percent moisture by volume range 0 – 20%.

When a special stainless steel mounting tube is used, the device is recommended for temperatures in excess of 2000°F.

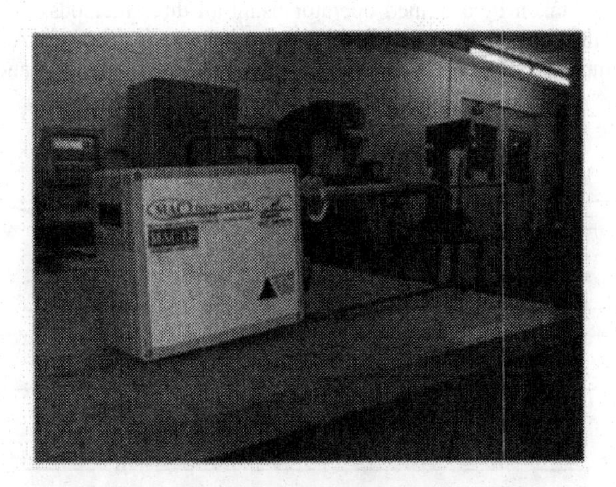

Figure 3 – Moisture Analyzer

Figure 4 – Moisture Analyzer Connected to Furnace Via Stainless Mounting Tube

The analyzer samples air and displays percent moisture by volume on a continuous basis, and adjusts to significant changes in moisture level within 60 seconds. It is accurate to ± 0.2%, which is the same level of accuracy that a Dräger tube provides. The unit can output a proportional voltage or current signal as well as to a digital display and so could be connected to a PLC or other device if desired. ESPC installed such a device on its production furnace in 2003 and has been using it for the last two years.

Comparison to Other Methods

The analyzer installed on the furnace at ESPC was compared to the Dräger tube and condensing mirror methods of moisture level measurement. Readings of the furnace moisture level were taken by a trained operator using all three methods. The readings were taken from the same furnace location at the same time and noted. A total of 20 readings were compared. The results are compared graphically in Figures 5 and 6.

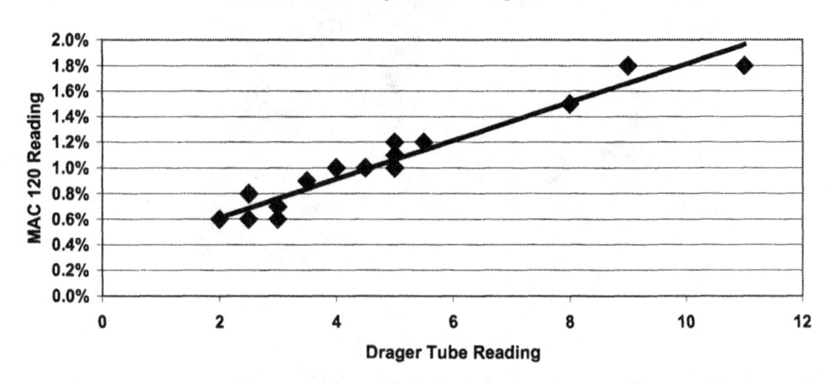

Figure 5 – Comparison of Analyzer Reading vs. Dräger Tube Reading

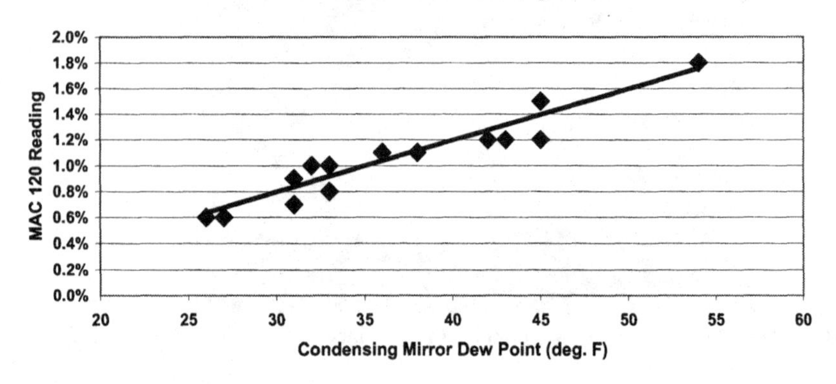

Figure 6 – Comparison of Analyzer Reading vs. Condensing Mirror Reading

The analyzer showed a good correlation to both other methods of moisture level measurement and was repeatable.

Results

The new moisture analyzer does work for measuring furnace moisture level on a continuous basis. It satisfies the three criteria set out earlier:

1) It is compatible with furnace environments at temperatures over 2000°F.
2) It monitors furnace moisture level on a continuous basis without direct operator involvement.
3) It gives repeatable results that correlate with the other two methods.

Control of Furnace Moisture Level

Control of furnace moisture level involves maintenance of the furnace and adjustments to deal with high or low moisture levels beyond the desired 1 – 3%. Maintenance and adjustment are both required to effectively control the furnace moisture level. Without proper control one is at the mercy of the ambient conditions.

Furnace Maintenance

To control the furnace moisture level, the furnace needs to be maintained. The following is a list of suggested items that should be maintained. A company that services furnaces can assist with these items if needed.

Air Seals – Ensure they are directed properly and are giving the proper air velocity. One wants to prevent hot air escaping but does not want cold air being sucked into the furnace.

Vents – Ensure they are open and unclogged.

Bisque – Ensure that the bisque is completely dried prior to the ware entering the furnace so as to reduce the amount of water introduced in this way.

Adjustments for Excessively High Moisture

High moisture is typically an issue in the summer, when ambient moisture levels are at their highest. High moisture can be dealt with by the injection of clean dry air into the furnace. The dry air helps to reduce the moisture level by absorbing some of the water and also helps to push moisture and any products of combustion in the furnace out through the vents. For this application, it is recommended that the air be supplied by a regenerative (desiccant) air dryer. This type of dryer provides air that is much dryer (dew points typically at -40°F or less) as compared to conventional dryers (dew points typically around 35°F).

The air should be injected at the top of the furnace in the middle of the firing area. The air should be introduced via large diameter tubes that are connected to a smaller diameter regulated air line. The large diameter tube allows the air to expand some as it enters the furnace, thus reducing its velocity. One should take care that the air is not introduced where it is blowing directly on the ware or stirring up dust in the furnace.

Adjustments for Excessively Low Moisture

Low moisture is typically an issue in the winter, when ambient moisture levels are low. Low moisture can be dealt with by the injection of steam into the furnace. A regulated steam line should be connected to the bottom of the furnace. The line should have a drain elbow installed for taking off any condensate before opening the steam line to the furnace.

Proper furnace maintenance and dry air/steam injection give one the ability to properly control the moisture level in the furnace. Control of the furnace moisture level helps to achieve a controlled, repeatable process. Process repeatability is critical in producing consistent high-quality products.

References

1. R. Ott. "Effect of Water Vapor in Furnace Atmosphere on Ground Coat Surface Properties"; Proceedings of the PEI, Volume 40 (1978), pp. 111 – 117.
2. R. Ott. "Effect of Furnace Moisture on Ground-Coat Surface Quality"; Proceedings of the PEI, Volume 44 (1982), pp. 370 – 373.
3. J. Cook. "Report of the PEI Furnace Atmosphere Committee"; Proceedings of the PEI, Volume 53 (1991), pp. 57 – 59.
4. PEI Technical Manual, Volume PEI-601, p. 18.
5. Machine Applications Corporation, Sandusky, OH. "The MAC Humidity/Moisture Handbook", pp. 1 – 21.

PREHEATED COMBUSTION AIR

Mike Horton
KMI Systems, Inc.

Abstract
Gas fired furnaces have been used for many decades in the enameling industry. Major improvements have occurred to enhance their efficiency in the structural designs, and burner designs. As the cost of fuel continues to be a significant part of the operational costs of enameling plants, fuel savings can be increased by the use of preheated combustion air. Exhaust air from the furnace can be used to preheat air going to the burners to achieve improved efficiencies as noted in slide No. 4 (Fuel Savings). As the preheated combustions air becomes hotter, the potential savings increase.

Combustion of (natural) gas releases great amounts of heat, water vapor and carbon dioxide. There are specific fuel-air ratios to achieve optimal combustion as shown in slide No. 6 (Fuel/ Air Ratio). To achieve the cost savings, recuperators are used to capture some of the exhaust heat energy as shown in slide No. 11 (Heat Exchanger). As with any machinery in an industrial setting, a control system is necessary to control and optimize the process. Savings possible with a well-engineered system may range from 17% to 22% of regular gas costs. However, the heat exchanger systems do require maintenance, may need repairs over time and have a range of performance variations depending on the general operation of the enameling furnace.

What is Preheated Combustion Air

- Exhaust air from the furnace is used to preheat the combustion air

GAS RATES

USD $ PER THERM OR 100 CUFT

Rate Class	'05	'04	'03	'02	'01	00	% DIF '00 TO '05
COMMERCIAL	.964	.917	.815	.593	.697	.539	+79%
INDUSTRIAL FIRM							
5x Demand	.964	.917	.815	.593	.697	.539	+79%
25 x Demand	.861	.821	.732	.546	.637	.457	+88%
Over 25 x Demand	.964	.917	.815	.593	.697	.442	+118%
INDUSTRIAL INTERUPTIBLE	.792	.756	.680	.513	.592	.414	+91%

Fuel Savings

COMBUSTION AIR TEMPERATURE °F

Furnace Gas Exit Temp. °F

% Fuel Saved	400	600	700	800	900	1000	1100	1200
700	7.9	12.1	-	-	-	-	-	-
800	8.2	12.4	14.5	-	-	-	-	-
900	8.4	12.8	14.9	16.8	-	-	-	-
1000	8.7	13.2	15.4	17.5	19.5	-	-	-
1100	8.9	13.7	15.7	18.0	20.0	22.0	-	-
1200	9.3	14.1	16.4	18.6	20.7	22.7	24.6	-
1300	9.6	14.6	17.0	19	21	23.4	25.4	27.2

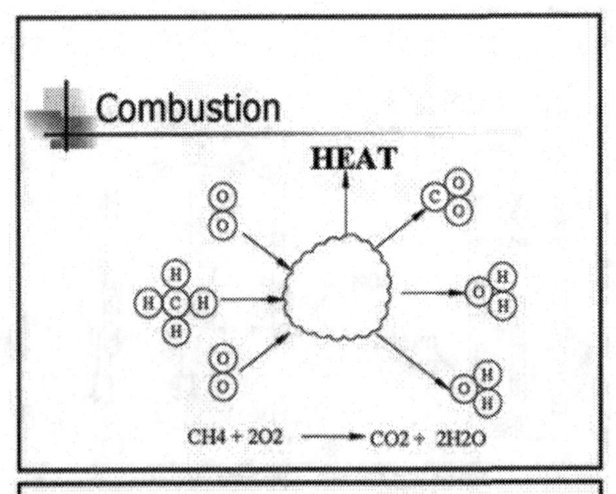

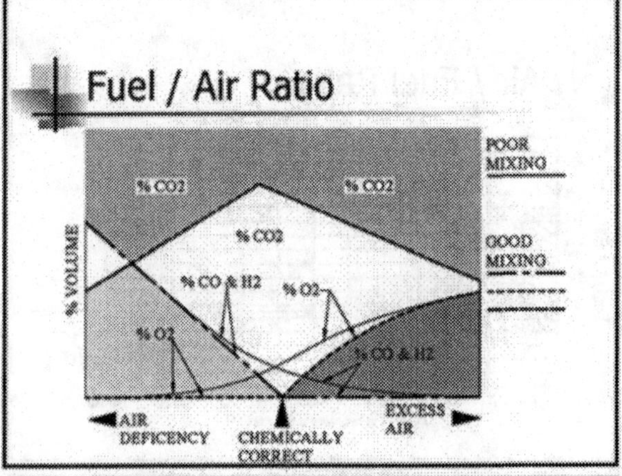

Benefits

- LOWER FUEL REQUIREMENTS
- COST SAVINGS FROM 17% TO 22%
- Example of savings for a 20,000 Lbs/Hour furnace

20,000 lbs/hour*420 Btu/LB* = **8,400,000 BTU/Hour**

8,400,000 Btu/hour * $8.75/1,000,000 BTU * 16 Hours/Day *243 Days/Year = **$285,768/Year**

Annual Savings@ 17% = $285,768/Year x 0.17 = **$48,580/Year**

Annual Savings@ 17% = $285,768/Year x 0.22 = **$62,870/Year**

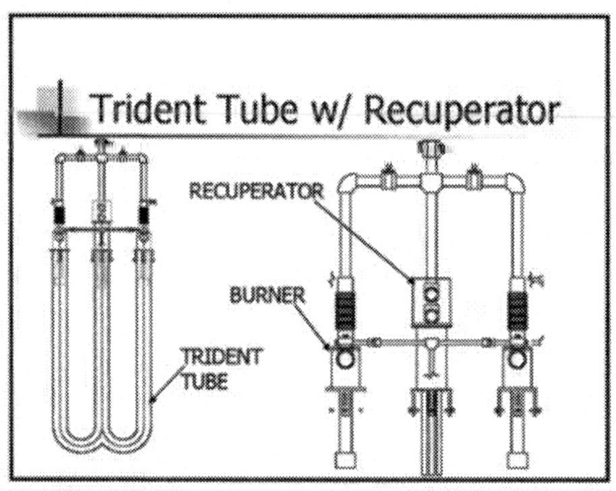
Trident Tube w/ Recuperator
RECUPERATOR
BURNER
TRIDENT TUBE

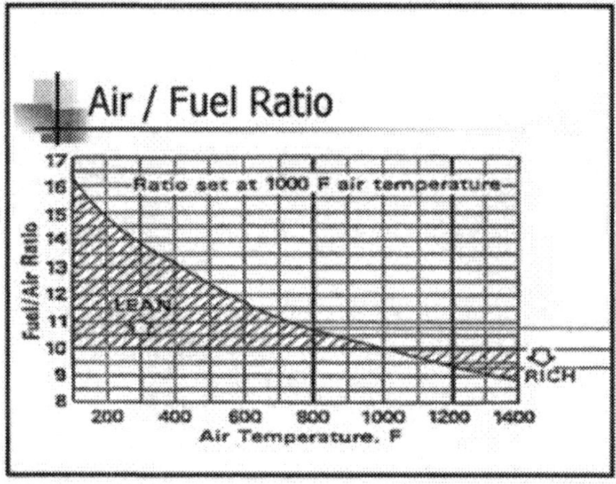
Air / Fuel Ratio
Ratio set at 1000 F air temperature
LEAN
RICH
Fuel/Air Ratio
17
16
15
14
13
12
11
10
9
8
200
400
600
800
1000
1200
1400
Air Temperature, F

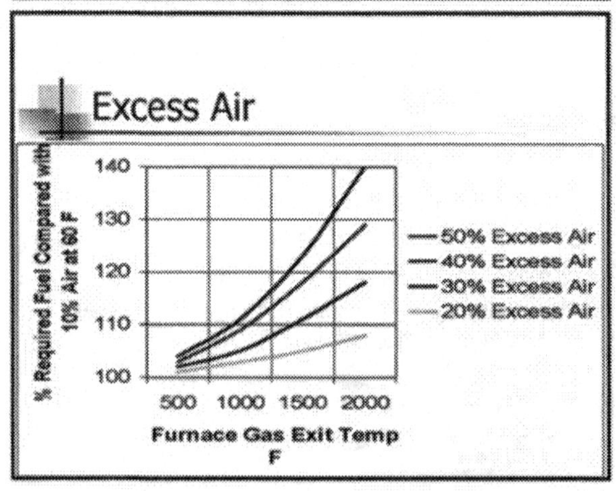
Excess Air
% Required Fuel Compared with 10% Air at 60 F
140
130
120
110
100
500
1000
1500
2000
Furnace Gas Exit Temp F
50% Excess Air
40% Excess Air
30% Excess Air
20% Excess Air

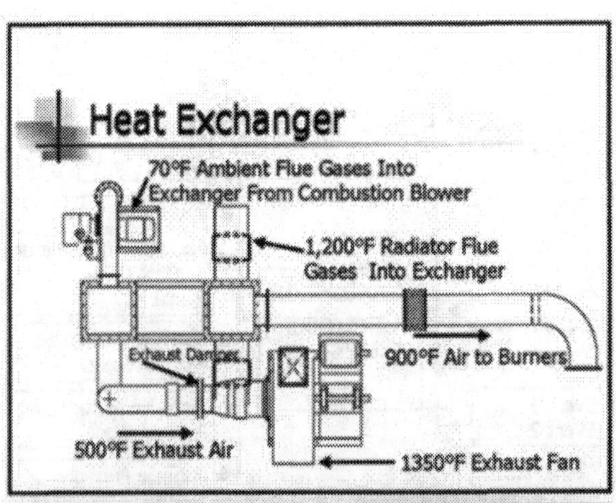
Heat Exchanger
70°F Ambient Flue Gases Into
Exchanger From Combustion Blower
1,200°F Radiator Flue
Gases Into Exchanger
900°F Air to Burners
Exhaust Damper
500°F Exhaust Air
1350°F Exhaust Fan

Typical Control Panel

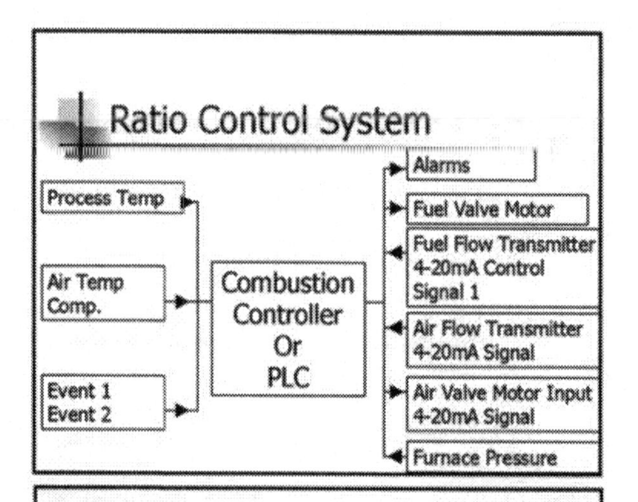

Advantages of Microprocessor Preheated Combustion Air Systems

- Fuel Air Ratio Control
- Effective Fuel Mixing
- Temperature Compensation of Preheated Air With Varying Burner Outputs
- Controller has Mixture Limits
- Component Reliability
- Rapid Scan Rates for Best Fuel Mixture
- Lowest Maintenance
- 17 to 22% Gas Savings

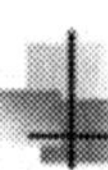

Disadvantages of Mechanical Preheated Air Systems

- Maintenance of Heat Exchanger
- Mechanical Types are set Lean to Obtain Ratio at Operation Temperatures
- Load Variations will Shift Fuel Mix from Lean to Rich
- Recuperators degrade and Leak
- High Pressure Combustion Air Leaks into the Low Pressure Gas Side

LOW TEMPERATURE CLEANERS

Mark Godlweski
Henkel Technologies

Abstract
Low temperature cleaners have been used in the enameling industry for many years. However, rapidly increasing energy costs continue to drive interest in low temperature systems in operations that may not have evaluated or use low temperature cleaners. This presentation outlines considerations for the use of low temperature cleaners. The types of soils typically associated with enameling are rust inhibiting oils and lubricants used in stamping of parts. The soils may be organic or inorganic. There are several types of natural and synthetic lubricants and each requires specific treatments for removal from a metal surface. Control of the parameters of the cleaning system is also important. Parameters which must be considered are temperature, time, percent of oils, concentration of cleaner, pH, and agitation of the cleaning system. Temperature control as "low temperature" does not mean the absence of temperature management. The low temperature cleaners will clean better at moderately elevated temperatures. However, too high a temperature will kick out surfactants and too low a temperature will cause foaming to occur.

In enameling operations, the cleaners are typically operated to allow overflow from the last tank to the previous tank to optimize the use of the solutions or rinses. Determining the cleanliness of the cleaned metal surface is somewhat subjective, however, the "water break" test is widely used followed by the "white glove" test. Both of these require observation; with the water break test to see if any water beading occurs or if the white glove has any visible stains on the cloth.

There are a variety of factors that influence cleaning and can be adjusted in an effort to maintain lower cleaning temperatures and overall lower operating costs. Some of these are types of soils (identification), cleaner time, cleaner concentration, agitation, cleaner make-up [chemistries] and finally the operator who is the most critical person on any cleaner line and is most responsible for a successful cleaning process.

Low Temperature Cleaners

Determination if a Low Temperature Cleaner will be feasible understanding of the following is imperative.

- Types of Soils/Contamination on Parts
- Reason for cleaning
- Cleaner Parameters
- Cleaner Mechanism/Chemistry
- Determination of a clean surface

Types of Soils

- Dust – grinding particulate
- Finger prints
- Rust inhibiting oils
- Waxes
- Mold release
- Coolants
- Lubricants
- Oxidation

Classification of Soils

There are two classifications of soils organic and inorganic.

Organic soils are oily, waxy films such as mill oils, rust inhibitors, coolants, lubricants, and drawing compounds. Alkaline cleaner are utilized to remove (clean) organic soils.

Inorganic soils are rust, smut, heat scale, laser scale, abrasives, flux and shop dust. Acidic cleaners are employed to remove inorganic soils.

Four Types of Metalworking Lubricants

1. Straight Oils
2. Emulsions
3. Semi-Synthetics
4. Synthetics

Straight Oils (Mineral Oils)

Used primarily for lubrication where there is little need for cooling. Subdivided into to groups Paraffinic oils (straight chain hydrocarbon bases) and Napthenic oils (ring structured hydrocarbon bases.

Paraffinic Structure

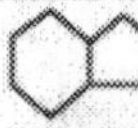

Naphthenic Structure

Paraffinic Oils

Utilized as engine oils due to their high viscosity and ability to keep their viscosity at elevated temperatures.

Semi-Synthetic (Micro-Emulsions)

Semi-Synthetic lubricants are more stable than emulsions lubricants because of the emulsion size (smaller oil globules). Typically contains mineral oil, emulsifiers, dispersants, boundary additives, anti-foaming agents, corrosion inhibitors, and biocides. More effective for cooling vs. lubricating.

Emulsion Oil Globules

0.01 microns → 0.02 microns

Synthetic (No Oils)

Synthetic lubricants are true solutions that contain ingredients that form very small emulsions, yet contain no oils. Synthetics are formulated from polyglycol, polyisobutylene, or poly alpha-olefin bases. Contain emulsifiers, amines, dispersants, corrosion inhibitors, lubricity additives, and anti-foaming agents.

Emulsion Micelle

0.015 microns → 0.005 microns

Control Parameters for Cleaners

Cleaning is accomplished with Alkaline Cleaners through the combined work of chemical, physical, and thermal application for a specific amount of time.

The following parameters allow the Cleaners to function properly:

- Temperature
- Concentration
- Time
- Agitation
- Percent oils
- Percent Surfactants
- pH.

Process Time

- Sufficient cleaning contact time is required to allow the cleaner to work.
- Cleaner times are fixed – thus temperature, concentration (increased alkalinity), and pressures may be adjusted to allow proper cleaning efficiency.

Temperature

- Increased temperature will aid is soil removal by softening the oily soils.
- Decreasing the interfacial surface tension, and allowing emulsification of oils.
- Temperature will be dependant upon the surfactant utilized
  - Too high temperature will kick out surfactant
  - Too low temperature will cause foaming to occur within the cleaner bath.

Chemistry – Concentration

- Free Alkalinity, which is a measure of the working component of the alkaline cleaner.
- Total Alkalinity, which is the total sum of hydroxide, carbonate, and bicarbonate components and will determine when to dump the alkaline cleaner bath

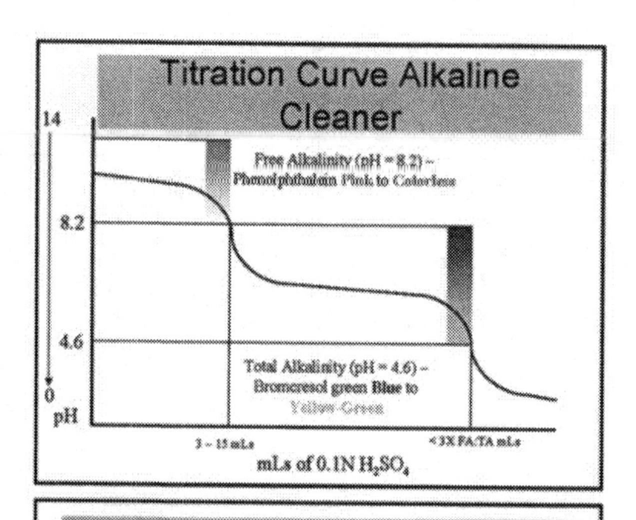

Agitation/Pressures

- Impingement within spray stages helps to physically and mechanically loosen up the soil allowing the cleaner to emulsify the oils
- Circulation within a immersion stage helps to get fresh cleaner within enclosed areas to allow the cleaner to emulsify the oils were a spray cleaner will be unable to reach (come in contact).

pH

- Alkalinity of the cleaner provides the high pH.
- At high pH oils are more readily emulsified enabling soil removal easier.
- High pH makes redepositing of oils more difficult.

Porcelain Enamel Cleaning Line

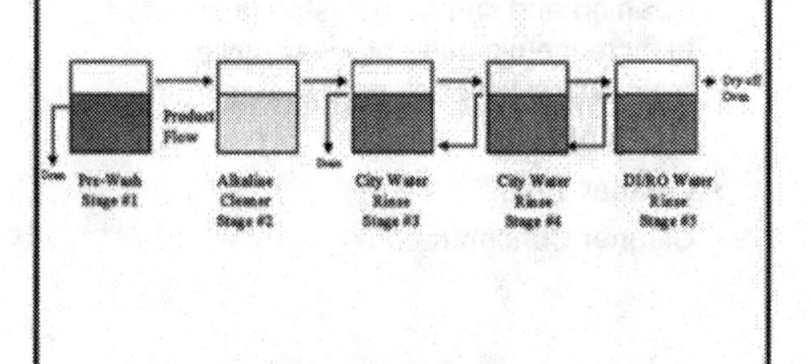

Determination of a Clean Surface

- There is no precise method to determine if the surface is clean, yet there are some methods that will help to evaluate surface cleanliness.
- One method is to evaluate a water-break-free surface. If contamination is present the water will bead up in un-cleaned areas.
- The second method would be a "white glove test" where the wet surface is wiped with a white paper towel and visually evaluate the paper towels for any colors (Black, Brown, and Red).

Water Break Free Test

"water break free" test. Water is sprayed onto the part, and the operator observes whether or not the applied water film runs off the part in the desired continuous sheeting action (clean), or is diverted in any areas around any areas which may still be contaminated with dirt, oily films, etc. (not clean).

Summary of Low Temperature Cleaning

There are a variety of factors that influence cleaning and can be adjusted in an effort to increase cleaning process while maintaining lower cleaner temperatures.

- Types of Soils
- Cleaner Time
- Cleaner Concentration

Summary of Low Temperature Cleaning

- Agitation
- Cleaner Chemistries
- Operator

  It is the operator who is the most critical person on any cleaner line and is most responsible for a successful cleaning process.

OPERATING ISSUES FOR SUCCESSFUL USE OF AMBIENT CLEANING SYSTEMS

Ken Kaluzny,
General Industrial Group, Coral Chemical Company

Abstract
I will start by briefly discussing the potential differences in ambient and heated cleaners. The operation of an ambient cleaner has a lot of similarities to heated cleaning. Both ambient and heated cleaners require solution control I will briefly discuss some solution control techniques. System maintenance is important regardless to whether the cleaner is heated or not. The same can be said for solution maintenance. I will conclude with discussing some inherent issues of ambient cleaning.

Cleaner Chemistry
Heated cleaners and ambient cleaners will contain the same builders. Hydroxides, carbonates, phosphates, silicates, borates and other inorganic salts are used in alkaline cleaners regardless of temperature. Sequestrates, such as gluconate or citrate, and chelates, such as EDTA and NTA, are used for water softening and inorganic soil removal.

The differences between ambient and heated cleaner chemistry are more evident in the additives used. Detergents or surfactants are different. Spray cleaner formulations contain nonionic surfactants. Nonionic surfactants are employed, as they are lower foaming than soaps and anionic surfactants. Nonionic surfactants have a property called cloud point. Cloud point is the temperature where the nonionic surfactant becomes more oil-like and floats to the surface or "clouds out." If you see this in a freshly charged tank the surfactant looks like clouds when the solution is idle. In dirtier systems it just becomes part of the oil layer. As the temperature moves towards the cloud point foam is reduced while cleaning is enhanced.

One reason for taking you through this chemistry lesson is that the ambient cleaner formulation is limited in surfactant selection. Low cloud point nonionic surfactants are sometimes used in heated spray cleaner products to act like a defoamer. As we are discussing ambient cleaners the surfactant choices will be less able to fulfill this role. Defoamers may be required in the formulation or used as tank side additions.

A second reason involves oil rejection. In heated formulations, low cloud point surfactants help split the oil from solution. Once again we are limited due to the ambient operation. Separate oil rejecting additives will be necessary to remove oil from the cleaning solution. Otherwise you will have to dump the cleaner bath more often.

Solvents may be used in the ambient cleaner. Solvents would be used to make up for the lack of heat. Specifically, soils that have a high pour point, that point at which they become fluid, will be difficult to clean at ambient temperatures. Solvents will help dissolve or soften the hard or waxy soil. This is particularly the case with some rust preventative oils and drawing compounds. Aging makes the soil less fluid through evaporation and accumulation of airborne particulate.

Heat vs. No Heat

I believe that it is important to know why something happens as it will provide you with chemical intuition. My esteemed competitor helped answer the question, "to heat, or not to heat." Here I want to point out some implications of operating an ambient system. The reduction of heat implies that there needs to be a rebalancing of TACT. Some of you may have heard of this put differently. Simply stated the acronym stands for Time, Activity, Concentration and Temperature. This fundamental view of a cleaning operation is commonly used to address cleaning issues. If you reduce one facet of TACT, you may need to increase another to facilitate cleaning. The elimination of heat puts everything else at a heightened sense of awareness. Increasing time is doubtful. Who slows down their line these days? Increased agitation or spray pressure is limited in that you may already have high action and you don't want parts to fall off their hooks. Increasing your chemical concentration is probably your best option. If you are dealing with easily cleaned soils, then you probably won't have to increase concentration. However if you do, then you will be looking at using more chemistry in the cleaner stage representing an increase in cleaner cost. The amount of alkali carried over to the rinse stages will be higher, which will effect wastewater treatment costs or challenge your discharge restrictions. Remember there is no such thing as a free lunch. The increase in chemistry and compliance costs will more than likely is less than the amount of energy savings.

Solution Control

Solution control is a very important element in pretreatment. You need to control the process if you want to control the out come of finished material. The cleaner concentration is generally controlled by wet chemistry methods. The free alkalinity titration is used to determine the concentration of an alkaline cleaner. Titrations are simply a measuring technique to determine the amount of cleaning product in the solution. A titration is performed by measuring a known amount of cleaner such as 10-milliliters and placing it into a beaker or flask. A color indicator is added to the same beaker. Titration solution is added to the beaker while the beaker is swirled to facilitate mixing. The titration solution has a known strength and is opposite in pH from the cleaner bath. 1.0 N and 0.1 N hydrochloric acid are commonly used for alkaline cleaners. The titration is complete when the color of the solution changes. In the case of alkaline cleaners, phenolphthalein is commonly used for the free alkalinity titration. The cleaner solution will turn red or pink when added to an alkaline cleaner. The addition of acidic titrating solution will slowly fade the color until all red and pink disappears. This is the endpoint. The amount of titration solution required for color change multiplied by a product factor will let you know how much cleaning product is in the solution.

Conductivity is commonly used with automated feed equipment. Conductivity will also measure by-products of the cleaning reactions as well as the accumulation of water salts. Since we are dealing with a system that won't have as much evaporative loss, there will be less concern in the effect of salt accumulation in an ambient cleaner than a heated cleaner. Although conductivity is not perfect, it can be used as an effective way to control chemical concentration when used in conjunction with the free alkalinity titration.

I put pH up on the slide because some of you may be using pH as a means to control cleaner additions. Believe who you want, but my advise to you is to stop doing it. You are wasting money. pH is a measurement of the acidic and alkaline constituents in a solution. It measures the balance. It does not measure ionic strength or rather how much is in the tank.

The total alkalinity titration is typically used to measure bath life or usefulness of the cleaning solution. As the cleaner bath ages and accumulates contaminants, the total alkalinity titration will increase in relationship to the free alkalinity titration. The ratio of total to free alkalinity is used to predict when the cleaner will loose effectiveness. There is no magic number for the total to free alkalinity ratio that tells you to dump and recharge the cleaner. The ratio is dependent on a particular cleaner with particular soils. If there are no changes and enough history, then the total to free alkalinity ratio will help determine the appropriate time to recharge the cleaner.

Oil splitting tests can also be performed to determine the cleaner's usefulness. The use of oil levels in determining the cleaner's usefulness is also dependent on consistency. An oil test is generally done by your cleaner vendor but can be done at your plant if you have the appropriate glassware, reagents, and a hot plate.

Solution Management

Solution management not only includes the solution control procedures but also includes the monitoring of spray pressure and operating temperature. In the case of ambient cleaning, you may think that temperature control is a non-issue. In the true sense of the word, ambient means that there will be no heat applied to the solution. However the temperature still should be managed. Consider the winter months or freshly charging the tank. The water could be as cold as 50° F. This is ambient temperature in the true sense of the word, but your cleaner may be formulated to work as low as 70° F. At 50° F you may have foaming problems or poor cleaning. There still has to be some temperature control. These physical parameters are important to the success and predictability of the system. Spray pressure and temperature need to be controlled within an established range.

Chemical additions can be made in a variety of ways. Manual additions are the most simplistic but involve the operator handling typically corrosive materials. Safety glasses, chemical goggles, face shields, and rubber aprons and gloves are typically used for handling corrosive materials. Appropriate personal protective equipment should be worn when handling chemicals. Although cleaners are not toxic they can damage human tissues.

Proportioning pumps can be used to continuously add chemical with a minimal of operator handling. The proportioning pumps add chemical using two controls: speed and stroke. The control knobs are graduated in percent. The speed control indicates how frequent the pump is energized. The stroke graduation controls the amount of chemical proportioned with each stroke. Proportioning pumps still require manual intervention, as the chemical concentration will vary as part configuration may vary. Parts with more soil or parts that drag out more cleaning solution will lower concentration faster than flat parts with little soil. Changes in part configuration and soil will require readjustment of speed and/or stroke. You also need to remember to turn them off at the end of the workday unless the pump is tied into the electricity of the conveyor.

Automatic control systems are the next best thing. Automatic controllers use proportioning pumps but also include equipment to monitor fluctuations in the chemical concentration. Conductivity is the most common parameter measured for the automatic control of a cleaner bath. A conductivity probe sends a signal to a control box. The control box is set with upper and lower conductivity set points to turn the pump off or on respectively. The use of automatic control systems will keep concentration more uniform to maintain consistent quality and cost. Automatic controls don't totally eliminate the need for an operator. The conductivity probe will need to be cleaned occasionally and standardized typically once/month.

Solution Maintenance

Contamination removal is a form of solution maintenance. Particulate filtration and oil separation will help prolong the cleaner's usefulness and reduce operating costs. Ultra-filtration can remove both particulate and oil although. Ultra-filtration should be used with the course removal systems such as bag or media filters and oil separators such as skimmers or coalescers to reduce membrane maintenance and replacement.

The dump cycle is another form of solution maintenance. Contamination removal systems can prolong the dump cycle, but eventually the bath will need to be dumped and recharged. Your quality and finish acceptance rate will determine the frequency of dumping your cleaner.

System Maintenance

The system needs routine maintenance to maintain quality and minimize operational cost. Nozzles should be inspected routinely. The exact frequency of inspection depends on your operation. If you have a problem with part sway on the hangers or parts falling off the hangers daily then you should check the nozzles daily. The inspector should observe nozzle direction. There should be a positioning strategy that focuses on the parts traveling through the spray zone. Flat spray nozzles are commonly used for cleaning. The nozzles should be aligned 15 – 20 degrees off of vertical. This will prevent sprays from interfering with each other. If there is a lot of spray exiting the front or the back end of the spray zone, then you should turn the nozzles slightly in towards the center of the spray zone.

The inspector should also look for plugged nozzles. The nozzles should be cleared so that a good spray pattern is maintained. It is not appropriate to poke the nozzle with a stick or welding rod, etc. This will damage plastic nozzles and ultimately the nozzle will plug again. It is better to disconnect the nozzles and remove the occlusion from the back of the nozzle. Plastic clip on nozzles are the easier to maintain than stainless steel or threaded nozzles.

The screens should also be checked and cleaned on a routine basis. The frequency of this procedure is also dependent on your particular situation. The screens need to be clear so that you don't burn out the pump through cavitation.

If automatic controllers are used, then the probes that are used to turn a pump on and off need attention. Conductivity probes typically need less attention than pH probes. Conductivity and pH probes require routine calibration. pH probes will require cleaning and recalibrating more often.

Contamination removal equipment will require attention as well. Oil separation units need to be maintained to retain their functionality. Particulate filters need to be cleaned or replaced. Ultra-filter membranes need to be cleaned and unfortunately replaced at times.

Descaling isn't necessarily a routine cleaning procedure. However it should be part of your system maintenance. Scale formation will be dependent on the cleaner chemistry, soil constituents, and water hardness. In ambient cleaners, scale formation will be less than with high temperature cleaners. If plate and frame heat exchangers are used to maintain a temperature range of 70 - 80° F, then they will need to be descaled at some time. Risers are the other concern. Scale build up will eventually break off in pieces and plug nozzles.

Inherent Issues

There are potential issues inherent in operating the ambient cleaning system. Once you have successfully implemented the ambient cleaner into your pretreatment system, you need to be aware of soil variation. The continual success of operating at ambient temperatures will be dependent on the consistency of the soils introduced into the system. Your operation may be soil sensitive. You need to be aware of the type of soils you will remove and need to know if there will be any changes.

There may or may not be foam concerns from the ambient cleaner that you use. However certain lubricants will cause foaming in your cleaner tank. Sometimes it is the lubricant alone that causes the foaming problem and other times the foam is created by reaction of the lubricant and cleaner chemistry. You should have defoamer near your line. Consider it insurance.

The cleaning mechanisms listed also occur in heated cleaning systems. Emulsification of oils, saponification of fats and dissolution of soils in general will be the mechanisms of cleaning. The significance to these mechanisms is that they don't split soil/oil very well, especially when there is no heat. You will need some alternative chemistry to enhance soil rejection. Otherwise you will need to dump your ambient cleaner bath more often than your heated cleaner bath. Remember there is no such thing as a free lunch.

Conclusion

The name of the game is control – control your washer or it will control you. Maintaining the cleaner's chemistry, contaminants, and equipment will make the operation of the pretreatment consistent and compliant.

UPDATE ON FUTURE ENVIRONMENTAL ISSUES THAT WILL IMPACT YOUR INDUSTRY FOR THE PEI ENVIRONMENTAL, SAFETY AND HEALTH COMMITTEE

Bill Nichols and Jack Waggener
URS Corporation

Abstract

The Occupational Safety and Health Agency (OSHA) has proposed new limits occupational exposure to hexavalent chrome (Cr^{+6}) in dusts, mists and fumes. The current permissible exposure limit (PEL) standard, 52 micrograms per cubic meter could be reduced to 1 microgram per cubic meter with an activation level of 0.5 micrograms per cubic meter. These proposed levels will impact porcelain enamel plants and suppliers of materials. The levels proposed are based on a "conservative model" with industry proposing 10 to 25 micrograms per cubic meter and unions suggesting less than 0.25 micrograms per cubic meter. The new lower levels are considered technically difficult to measure. Final rules after comments are to be published in January 2006 if there are no delays.

Additional metals that may be impacted in the next few years are nickel and cobalt in drinking water as well as lead and other metals under the "World Summit on Sustainable Development" which the European Union and the EPA are key participants. The goal is to limit the use of toxic metals in materials and products and encourage recycling to reduce disposal issues. This will certainly impact the enameling industry and stainless steel products for cooking in the years ahead.

OSHA Proposed Rule:

Occupational Exposure to

HEXAVALENT CHROME (Cr+6)

In

Dust, Mists & Fumes

Permissible Exposure Limit (PEL)
(8-hr TWA)

Current PEL	**52 ug/m³**
Proposed PEL	**1 ug/m³**
Proposed Action Level	**0.5 ug/m³**

Extremely Low PEL
Impacts Many Processes in P.E. Plants
& Suppliers

Cr Oxides (PE)
Welding SS
Cr Sealers (Paint Lines)
Cr Electroplating
Welding Mild Steel
Chromates
Polishing/Grinding
Anodizing

Regulatory Schedule

Proposed	Oct. 4, 2004
Comments Due	Jan. 3, 2005
Public Hearing	Feb. 1, 2005

- 12 days/longest ever
- Industry
- Unions
- OSHA

Final Rule	Jan. 18, 2006

PEL
Health Based for Lung Cancer

OSHA "1.0 ug/m³

(based on conservative model)

Industry 10 to 25 ug/m³

(model plus no anecdotal facts)

Unions "less than 0.25 ug/m³"

PEL Needs to be Technically Feasible

1) Difficult to Measure
2) Engineering Controls
   Local Ventilation
   Fume Suppressants
3) Substitutes
4) Lastly, respirators
   (could be air supplied)

Economically Feasible

OSHA: $250 million/yr.
Industry: $4.0 Billion/yr.

Result:
- Closure of Small Business Surface Finishers
- More facilities move out of USA

PEI working to Help You!

EPA – on the horizon

Nickel & Cobalt
- Developing Drinking Water Standard
  2 yr.+ process
- Could Impact Exposed Surfaces of Ground
  Coats & other PE
    Water Heaters
    Cookware
- Impact Stainless Steel

European Union/EPA

"World Summit on Sustainable Development"

Limiting Use of Toxic Metals in Materials, Products

Goal: Reduce Disposal Issues & Encourage Recycling

Timeframe: Now and many years

Some Targets:

Lead	Nickel
Chrome	Cobalt
Many More	

Will Impact Materials & Manufacturing Decisions in World Market

PORCELAIN ENAMELS WITH A METALLIC APPEARANCE

Charles Baldwin and Louis Gazo
Ferro Corporation, Cleveland, OH

Introduction

As a fused glass coating, porcelain enamel has long been used to protect metal substrates from chemical attack, abrasion, and other corrosive and mechanical damage while also serving a decorative function. While a wide color and gloss range has been possible using traditional oxide pigments and other additives, it has not been possible to formulate enamel with an appearance similar to uncoated aluminum or stainless steel. Unfortunately for frit suppliers, this has left stainless steel as a major viable alternative to an enameled surface.

Generally, the only vitreous metallic finishes have been achieved using pearlescent pigments. These are flakes of mica coated with titanium dioxide or some other type of surface modification. While these can provide a sparkle appearance, they do not have the opacity of metallic pigments.

Organic fluoropolymer and silicone coatings with a metallic appearance are commercially available, but, relative to porcelain enamels, these have poor mechanical and thermal properties. Using new additive technology, enamels with a true metallic appearance have been developed that can be applied to aluminum, aluminized steel, or steel. These have the performance of vitreous enamel while having an aesthetically pleasing bright metallic finish. The metallic look can be used to break up the uniform shades of black and white that have been very common on porcelainized surfaces.

Metallic Pigments

The perceived color of metal arises from the wavelength distribution of incident radiation. Metals are opaque because incident radiation within the range of visible light excites electrons into unoccupied states above the Fermi energy. The energy of a photon then emitted by an electron as it moves from a high to low energy state equals that of the original absorbed electron. A bright silvery appearance with incident white light shows that the metal is highly reflective over the entire visible spectrum [1]. Because of the band gap in the electronic structure of oxide pigments and glasses, the energy of the emitted photons does not necessarily equal that of absorbed photons. Thus, the oxides are colored and a silvery metallic appearance is difficult to obtain in practice.

Metallic pigments are a small, useful class of inorganic colorants made up of fine particles of ductile metals. Metal pigments are especially good for providing a metallic appearance because the opacity, metallic color, sheen, and density of the bulk metal is retained [2].

Metallic pigments originated with the beaten gold foil used decoratively by the Egyptians alongside the first examples of porcelain enamel. Over the years, the foil was beaten thinner and thinner and eventually into gold powder used in artwork and inks. The high cost of gold lead to a search for substitutes. In the nineteenth century,

Sir Henry Bessemer invented a mechanical stamping process using steel hammers on steel anvils to make metal foils. At the same time, advances in aluminum smelting decreased the cost of aluminum. The explosion risk associated with dry communitation of aluminum was solved in the 1920s with wet ball milling. Since the 1970s, there has been an upswing in the use of metallic products. New finishes with a metallic flair are expected to be in demand on automobiles through 2007 – 2008 [3].

Metal	Color	Brilliance	Density (g/cc)	Corrosion Resistance	T_m (°C) [4]
Aluminum	White	High	2.7	Moderate	646 – 657
Nickel	Off-white	Medium	8.9	Good	1454
Stainless Steel	Off-white	Medium	~8.0	Good	1370 - 1400

Table 1. Properties of metallic pigments

Table 1 shows selected properties of three types of metallic pigments. The most common metallic worldwide, aluminum pigments have long been used in paints and inks. Aluminum is particularly attractive because of the silvery color, low density, and relatively low cost. However, to be used in aqueous coatings, considerable work has been invested in the development of encapsulation technologies to prevent corrosion and outgassing caused by the water and alkaline pH values [5]. Nickel provides a slightly yellow color with a rich luster. For stainless steel, alloy 316 is the most commonly used, and stainless steel pigments offer advantages over aluminum pigments. When added to an epoxy coating, an increase in the abrasion and corrosion resistance was observed in addition to the decorative effect [6]. Several other pigments including aluminum-bronze, copper, zinc, iron, silver, and titanium have been used to make metallic organic coatings.

The two main types of metallic pigments are leafing and non-leafing grades. Leafing pigments align horizontally at the coating surface to produce a dense, scale-like layer. Non-leafing pigments do not orient at the surface and create a random, "sparkle" effect. Specifically, sparkle is the reflection of light in a non-uniform manner. Whether or not a metallic colorant will leaf depends on the hydrophobocity of the fatty acid with which it was milled by the pigment supplier [7].

Metallic Enamel Formulation

Metallic enamel was made by blending commercially available metallic pigments with a low-temperature enamel system. To coat steel parts, a low-temperature frit with a linear thermal expansion of about 14 x 10^{-6}/°C was milled in a pigment-free formulation suitable for application to aluminized steel. The mill was emptied, a proprietary wetting agent was added, and a few percent of the metallic pigments was blended-in. The basic recipe is shown in Table 2. To coat aluminum parts, a clear enamel with a suitable low-temperature, higher-expansion, about 16 x 10^{-6}/°C, frit was used.

Raw Material	Amount
Enamel Slip	100
Pearlescent Pigment	0 – 5%
Metallic Pigment	0 – 5%
Wetting Agent	5 drops/L enamel slip

Table 2. Metallic enamel formulation

Steel parts such as those shown in Figure 1 were then enameled by first applying an electrostatic base-coat typically used for 2C1F applications. This was fired , the metallic enamel was applied using wet spray methods, and then the part was fired at 538°C for sufficient time for the enamel to show gloss. The drip pan, lantern hood, and grill all showed a silvery, metallic appearance not unlike bare stainless steel.

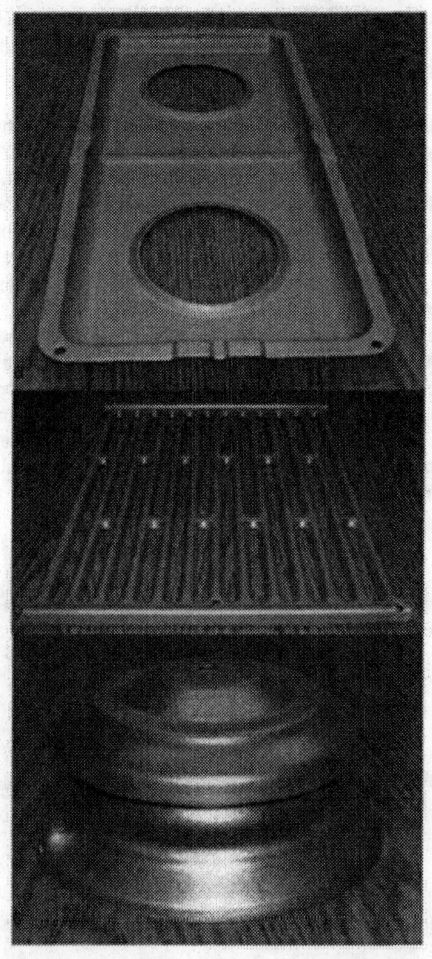

Figure 1. Three examples of parts coated with metallic enamel: (a) drip pan, (b) grill, (c) lantern hood

Testing

The acid reistance of the metallic enamels was determined by the frit used in the clear enamel milling. The spot acid resisance was tested using ASTM C 282-99 "Standard Test Method for Acid Resistance of Porcelain Enamels (Citric Acid Spot Test)." This procedure is equivalent to PEI T-21 "Test for Acid Resistance of Porcelain Enamels (Citric Acid Spot Test)." Several drops of a 10% aqueous solution of anhydrous citric acid were placed on the panel underneath a watchglass for 15 minutes. The degree of etching was rated from C to AA with AA being the best possible rating. When the medium expansion frit was used, an acid resistance rating of A was obtained.

To verify that the low-temperature enamel was well bonded to the ground coat, a drip pan like the one shown in Figure 1 was thermally shocked. The part was held at 316°C for 30 minutes and then immersed in room temperature 21°C water. After three cycles, none of the metallic coating spalled or flaked off the part, indicating good bond, which was then confirmed with impact testing.

The abrasion resistance of the metallic enamels was compared to organic coatings and uncoated stainless steel using the taber abrasion test described in ASTM D 4060-95 "Standard Test Method for Abrasion Resistance of Organic Coatings by the Taber Abraser." For this test, 10 cm by 10 cm square panels with center holes were abraded on a Taber abraser. The panel rotated on the abrasion machine for a set number of cycles under weighted abrasive wheels. Every 1000 cycles, the wheels were re-surfaced using silicon carbide paper disks, and a total of 2000 cycles were run. All of the panels were tested with the most abrasive CS-17 wheels. The panels were weighed before and after the test. Metallic coatings made with either stainless steel pigment or aluminum pigment were compared to conventional aluminum enamel, bare stainless steel, and two commercially available high-temperature organic coatings. The organics were a silicone-polyester used on griddles and bakeware and a PTFE-type coating. The PTFE coating was from a high-end piece of cookware and was a three-layer coating with a base coat, a filled interlayer for abrasion resistance, and a fluoropolymer-rich surface layer for hydrophobocity.

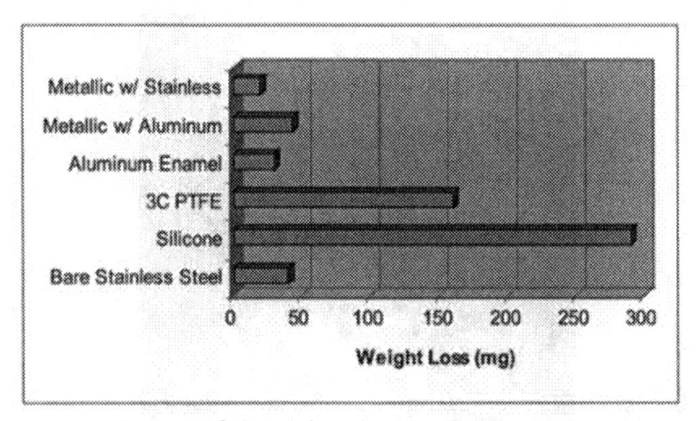

Figure 2. Comparative taber abrasion test results

The taber abrasion results are shown in Figure 2. Because it is not thought to contain re-reinforcement, the silicone-polyester showed the most weight loss followed by the PTFE coating. The metallic enamels showed similar weight loss to a traditional aluminum enamel and to bare stainless steel.

The scratch resistance of the coatings was evaluated using ASTM D 3363-00 "Standard Test Method for Film Hardness by the Pencil Test." The following hardness scale is assigned to a set of drawing leads where 6B is the softest and 9H the hardest:

6B – 5B – 4B – 3B – 2B – B – HB- F – H – 2H – 3H – 4H – 5H – 6H – 7H – 8H – 9H

The test plate was placed on a firm, level horizontal surface. Starting with the hardest lead, the pencil was held firmly against the panel at a 45° angle and pushed firmly away from the test operator. Sufficient force was exerted to either crumble the lead or to cut through the coating. This was repeated going down the scale until a pencil was found that would not cut through the film or scratch the surface for a distance of at least 3 mm.

Coating	Pencil Hardness
Bare Stainless Steel	5H
Silicone	HB
3C PTFE	7H
Aluminum Enamel	> 9H
Metallic w/ Aluminum	> 9H
Metallic w/ Stainless	> 9H

Table 3. Pencil hardness of coatings

Table 3 shows the hardest pencils that would not rupture or scratch the surface. The metallic enamels and conventional aluminum enamel could not be scratched with the hardest 9H lead. Because of the reinforcement, the PTFE had a hardness of 7H. Stainless steel could be scratched with a 5H pencil, and the silicone-polyester was the softest and could be scratched with an HB lead.

The heat resistance of the metallic enamels was compared to silicone-polyester paints by exposing coated panels to 400°C for two 50-hour intervals. The LAB color was measured initially, at 50 hours, and at 100 hours. The color change ΔE was calculated using Equation 1.

$$\Delta E = \sqrt{(L_i - L_f)^2 + (a_i - a_f)^2 + (b_i - b_f)^2} \qquad \textbf{Equation 1.}$$

In Equation 1, L_i, a_i, and b_i are the LAB color at 0 hours and L_f, a_f, and b_f are the LAB color at 50 or 100 hours.

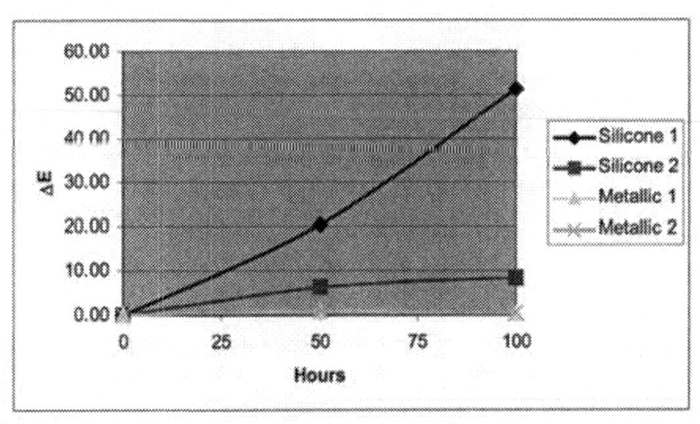

Figure 3. Change in color versus hours at 400°C

Figure 3 shows ΔE versus hours at 400°C for two silicone-polyester coatings and two metallic aluminum enamels. The first silicone completely changed color. Even though the second was colored with an inorganic oxide, it still significantly changed color and was severely degraded after 100 hours. The metallic enamels showed very low values for ΔE and retained their original appearance after 100 hours at 400°C.

The LAB color of a stainless steel 304-alloy coupon was measured before and after exposure to 400°C for one hour. Initially, L = 51.09, a = -0.14, and b = 0.51. Afterwards, L = 40.24, a = 3.10, and b = 7.14 with ΔE = 13.12. With $\Delta E < 1$ after 100 hours at 400°C, the metallic enamels showed superior color stability when exposed to heat.

Summary

Metallic enamels were made by blending metallic pigments used in the paint industry into clear enamel systems. Aluminum, aluminized steel, and enameling steel were coated. The metallic enamels were more abrasion and scratch resistant than the most heat-resistant organics currently used to make metallic paint. The color and gloss stability of two of the metallic enamels was superior to a silicone-polyester coating and to bare stainless steel. Therefore, metallic enamels can potentially replace bare stainless steel surfaces or paints to offer a more durable, longer-lasting finish.

References

[1] William D. Callister, Jr., *Materials Science and Engineering: An Introduction*, New York: Wiley, 1990, 713 - 4.

[2] Ian Wheeler, *Metallic Pigments in Polymers*, Shawbury, UK: Rapra Technology Limited: 1999, 3.

[3] Christine Canning Esposito, "Metallic Pigments: New Finishes on the Rise"
Coatings World 2002.
http://www.coatingsworld.com/May022.htm
(25 September 2003).

[4] William D. Callister, Jr., *Materials Science and Engineering: An Introduction*, New York: Wiley, 1990, 738 - 9.

[5] Bodo Müller, "Polymeric Corrosion Inhibitors for Aluminum Pigment," *Reactive & Functional Polymers* **39** (1999) 165 – 177.

[6] M. Selvaraj, "Stainless Steel Powder as a Protective and Decorative Pigment for Steel Structures in Organic Coating Industries," *Anti-Corrosion Methods and Materials* 44 (1997) 13 – 19.

[7] Martha Davies, "Understanding Aluminum Pigments," *Industrial Paint & Powder* 76 (2000) 38 – 40.

TWO COAT – ONE FIRE WET OVER WET SYSTEMS

Sebastien Humez
Pemco International

Abstract
Combining more than one operation at a time such as two coat – one fire increases production capacity and reduces manufacturing costs. The utilization of two coat – one fire wet over wet from the traditional two coat – two fire system is low investment, no additional technical expertise needed and is flexible regarding the type of steels that may be used. An overview of the process shows that the following steps are followed: washer → dryer → ground coat spray → cover coat spray → dryer → furnace. During the firing process, various reactions occur in the enamel as shown in figure No. 10. Initially various gases, water vapor and carbon dioxide are evolved from the attached water molecules as well as rapid oxidation of the underlying steel substrate. Once the enamel layers fuse to a continuous film, external oxidation ceases and various reactions occur at the steel-enamel interface. Correct timing of the various reactions is critical for optimum results. This process can be a great source of savings if the application parameters are closely controlled along with the enamel formulations.

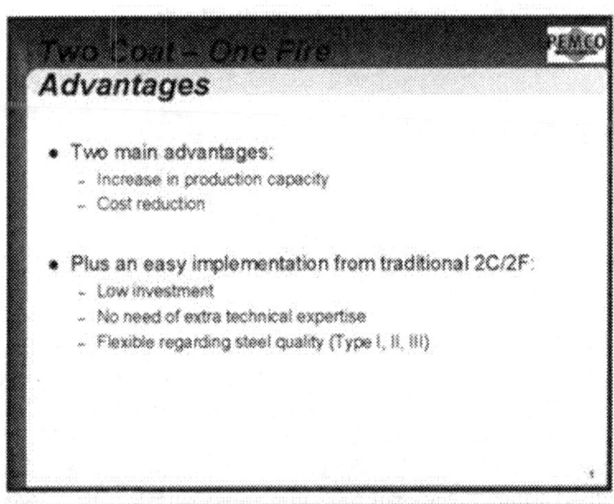
Two Coat – One Fire
Advantages
PEMCO
• Two main advantages:
– Increase in production capacity
– Cost reduction
• Plus an easy implementation from traditional 2C/2F:
– Low investment
– No need of extra technical expertise
– Flexible regarding steel quality (Type I, II, III)
1

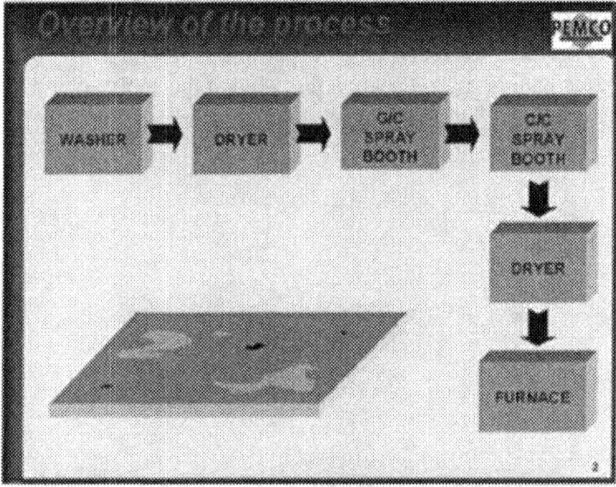
Overview of the process
PEMCO
WASHER
DRYER
G/C
SPRAY
BOOTH
C/C
SPRAY
BOOTH
DRYER
FURNACE
2

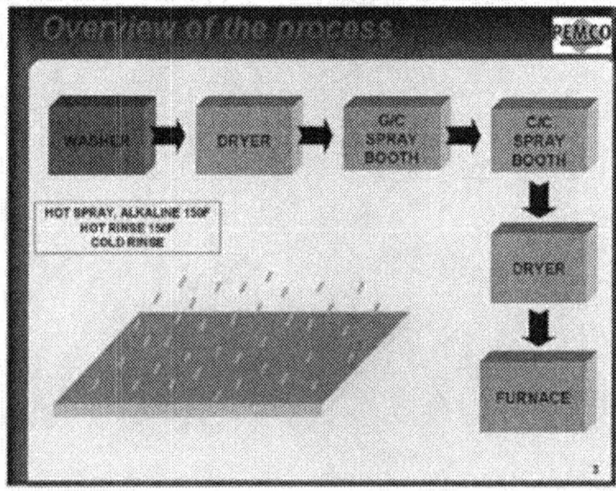
Overview of the process
PEMCO
WASHER
DRYER
G/C
SPRAY
BOOTH
C/C
SPRAY
BOOTH
HOT SPRAY, ALKALINE 150F
HOT RINSE 150F
COLD RINSE
DRYER
FURNACE
3

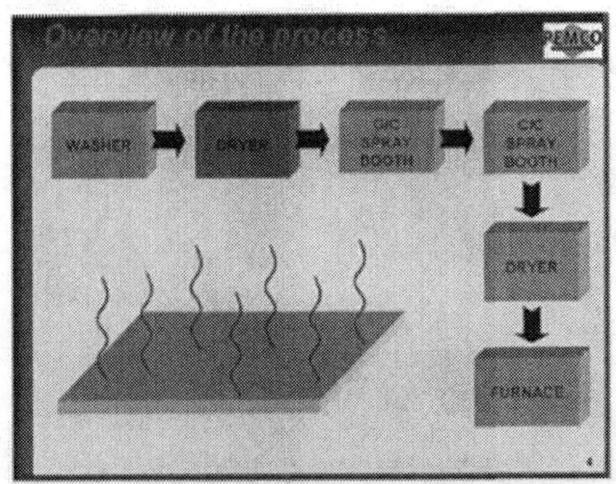
Overview of the process
PEMCO
WASHER
DRYER
G/C SPRAY BOOTH
C/C SPRAY BOOTH
DRYER
FURNACE

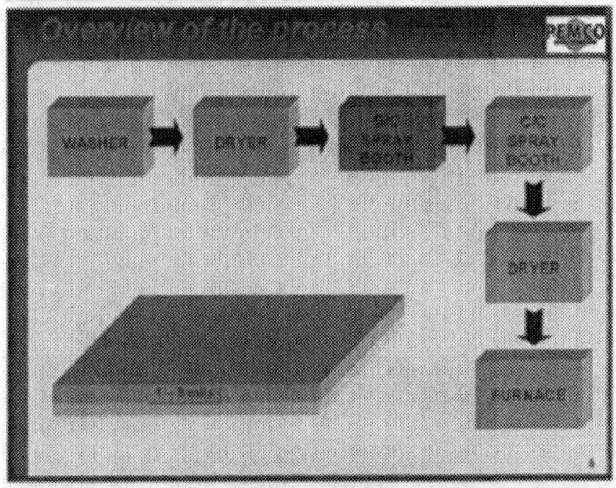
Overview of the process
PEMCO
WASHER
DRYER
G/C SPRAY BOOTH
C/C SPRAY BOOTH
DRYER
FURNACE
1 – 3 mils

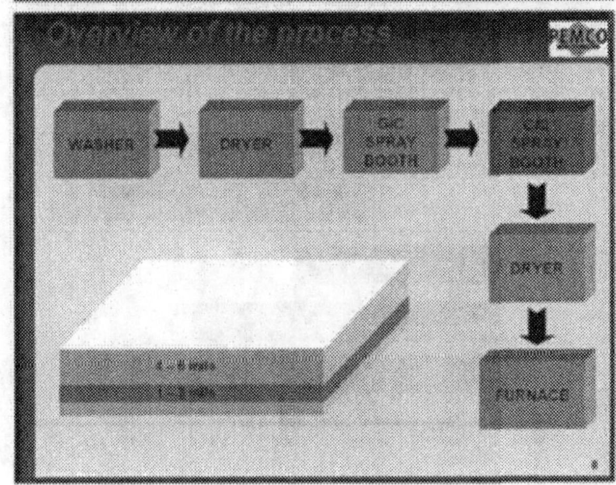
Overview of the process
PEMCO
WASHER
DRYER
G/C SPRAY BOOTH
C/C SPRAY BOOTH
DRYER
FURNACE
4 – 6 mils
1 – 3 mils

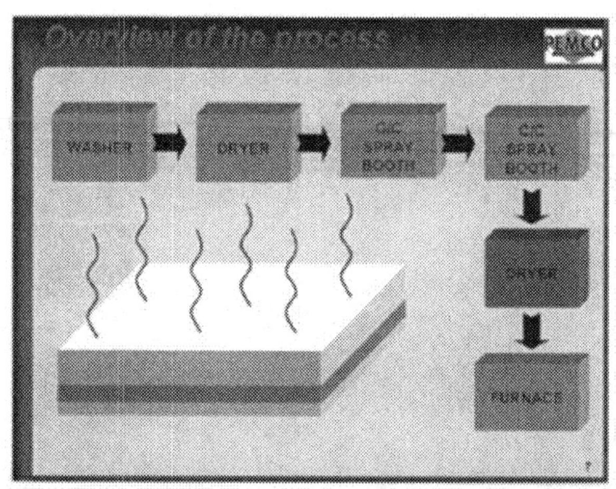
Overview of the process
PEMCO
WASHER
DRYER
C/C SPRAY BOOTH
C/C SPRAY BOOTH
DRYER
FURNACE

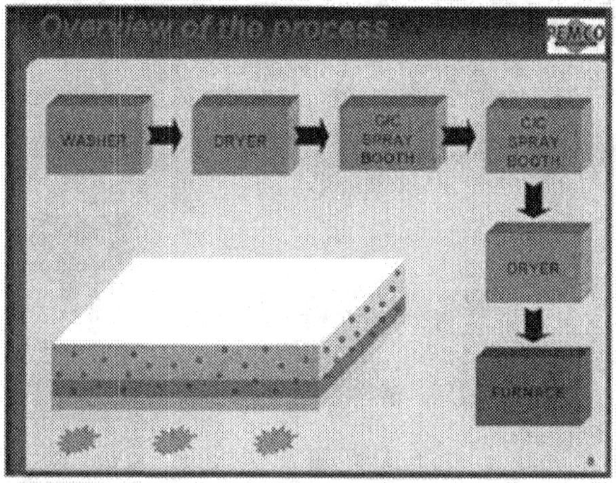
Overview of the process
PEMCO
WASHER
DRYER
C/C SPRAY BOOTH
C/C SPRAY BOOTH
DRYER
FURNACE

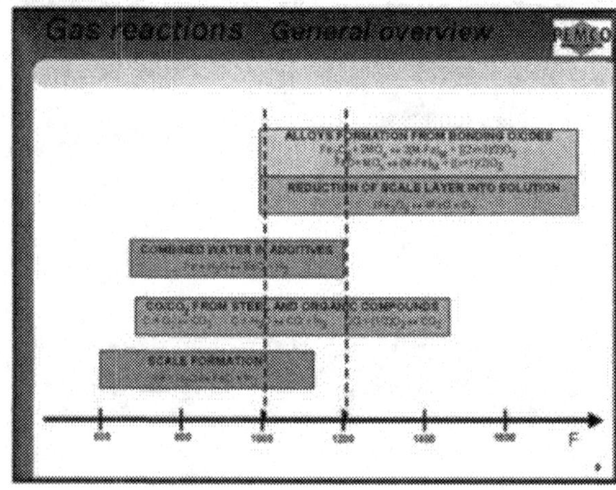
Gas reactions General overview
PEMCO
ALLOYS FORMATION FROM BONDING OXIDES
REDUCTION OF SCALE LAYER INTO SOLUTION
COMBINED WATER IN ADDITIVES
SCALE FORMATION
F

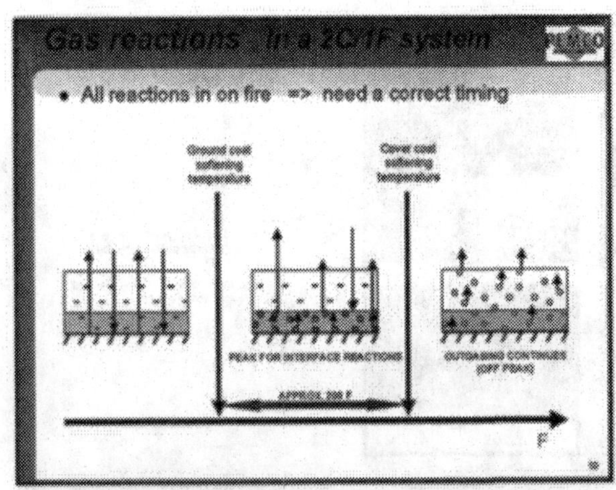
Gas reactions - in a 2C/1F system
All reactions in on fire => need a correct timing
Ground coat softening temperature
Cover coat softening temperature
PEAK FOR INTERFACE REACTIONS
OUTGASSING CONTINUES (OFF PEAK)
F

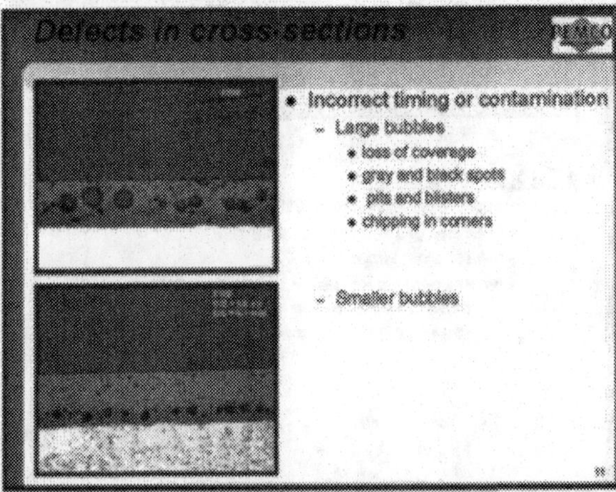
Defects in cross-sections
Incorrect timing or contamination
Large bubbles
loss of coverage
gray and black spots
pits and blisters
chipping in corners
Smaller bubbles

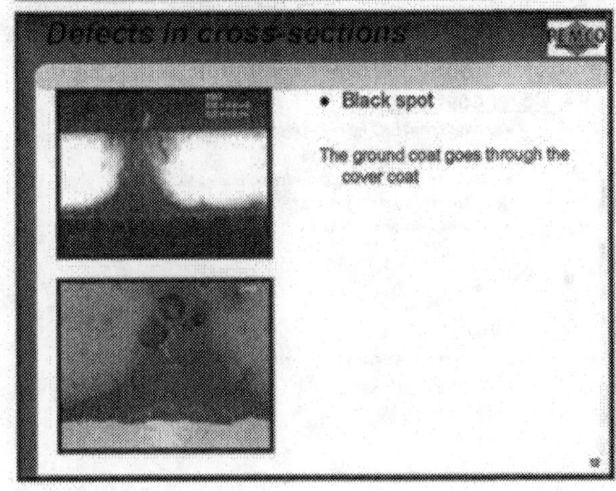
Defects in cross-sections
Black spot
The ground coat goes through the cover coat

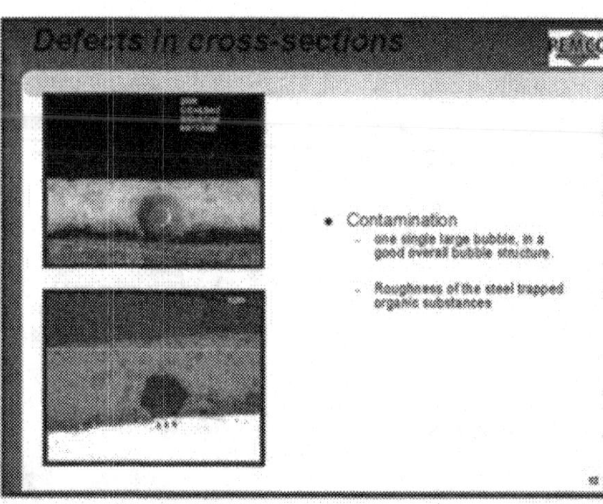
Defects in cross-sections
PEMCO
Contamination
one single large bubble, in a good overall bubble structure.
Roughness of the steel trapped organic substances

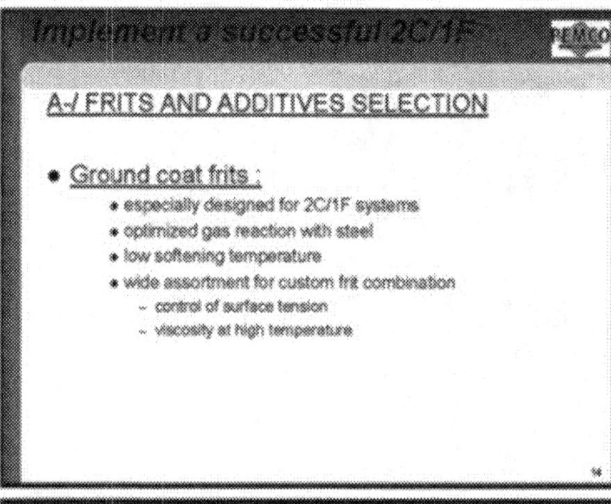
Implement a successful 2C/1F
PEMCO
A-/ FRITS AND ADDITIVES SELECTION
Ground coat frits :
especially designed for 2C/1F systems
optimized gas reaction with steel
low softening temperature
wide assortment for custom frit combination
control of surface tension
viscosity at high temperature
14

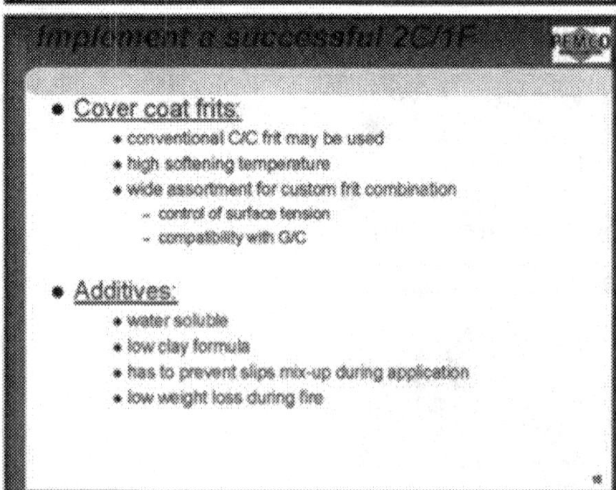
Implement a successful 2C/1F
PEMCO
Cover coat frits:
conventional C/C frit may be used
high softening temperature
wide assortment for custom frit combination
control of surface tension
compatibility with G/C
Additives:
water soluble
low clay formula
has to prevent slips mix-up during application
low weight loss during fire

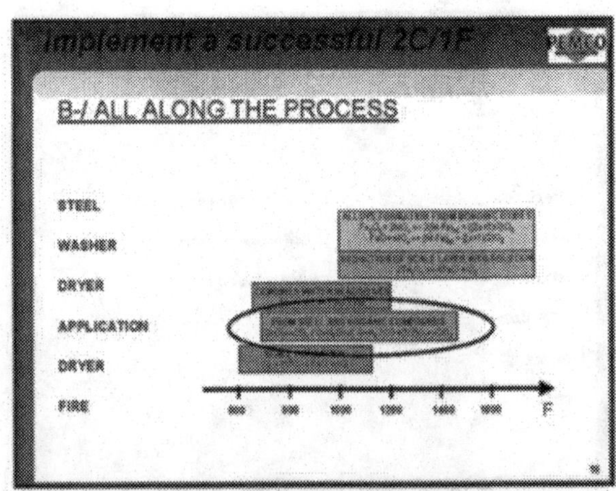
implement a successful 2C/1F
PEMCO
B-/ ALL ALONG THE PROCESS
STEEL
WASHER
DRYER
APPLICATION
DRYER
FIRE
F

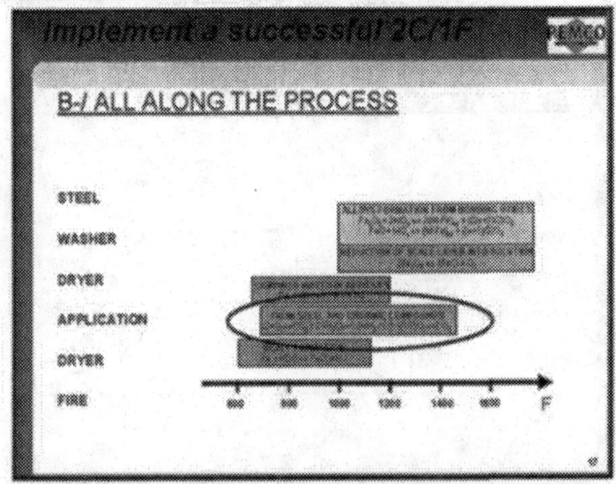
implement a successful 2C/1F
PEMCO
B-/ ALL ALONG THE PROCESS
STEEL
WASHER
DRYER
APPLICATION
DRYER
FIRE
F

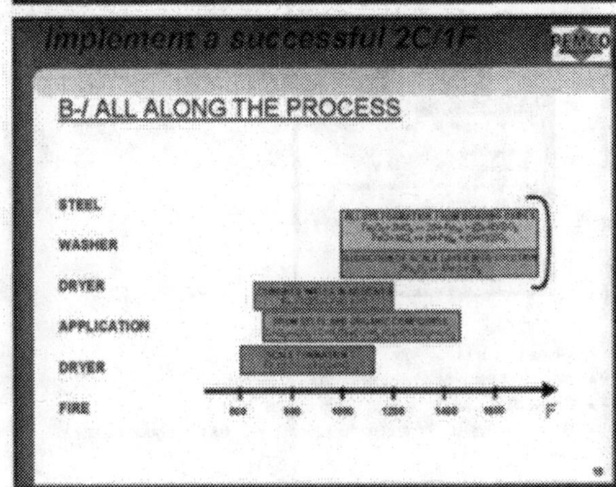
implement a successful 2C/1F
PEMCO
B-/ ALL ALONG THE PROCESS
STEEL
WASHER
DRYER
APPLICATION
DRYER
FIRE
F

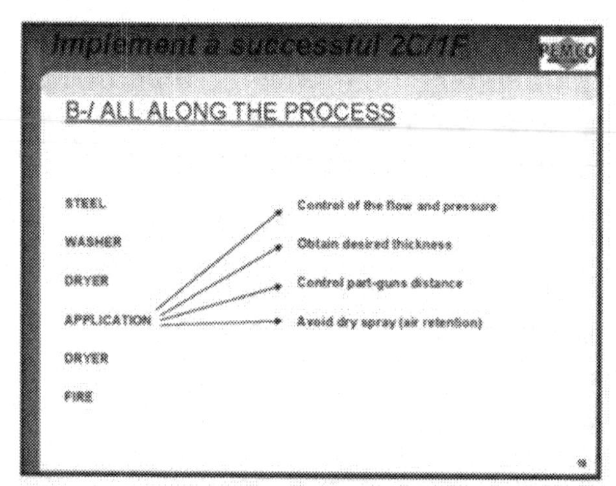
Implement a successful 2C/1F
B-/ ALL ALONG THE PROCESS
STEEL
WASHER
DRYER
APPLICATION
DRYER
FIRE
Control of the flow and pressure
Obtain desired thickness
Control part-guns distance
Avoid dry spray (air retention)

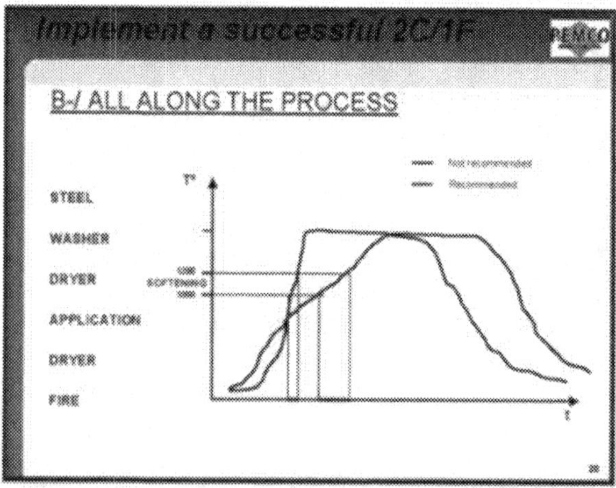
Implement a successful 2C/1F
B-/ ALL ALONG THE PROCESS
STEEL
WASHER
DRYER
APPLICATION
DRYER
FIRE
SOFTENING

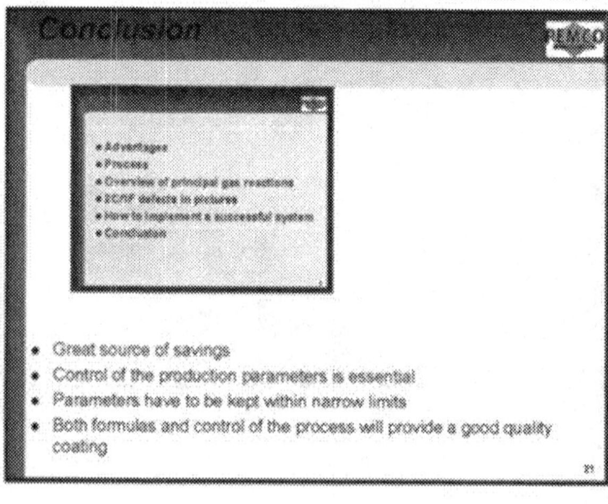
Conclusion
• Advantages
• Process
• Overview of principal gas reactions
• 2C/1F defects in pictures
• How to implement a successful system
• Conclusion
• Great source of savings
• Control of the production parameters is essential
• Parameters have to be kept within narrow limits
• Both formulas and control of the process will provide a good quality coating

TECHNOLOGICAL CHANGES IN PORCELAIN ENAMELING, 1955 – 2005

William D. Faust
Ferro Corporation, Cleveland, Ohio

Abstract
Significant changes have occurred the enameling industry over the last half- century. Various technical aspects and tests related to porcelain enamels are reviewed and discussed in light of today's needs.

Introduction
The enameling industry has made significant changes in the last 50 years. The processing of porcelain enamels was very dependent on the metal preparation techniques and the skill of the individuals doing the enameling. Reviewing the porcelain enamel literature since 1955, many transitions have occurred with a great deal of dedicated effort to adapt to the technological challenges.

1955 to 1965
This decade saw the development of aluminum enameling, furnace profiling with an early "Bozin Box", one coat enameling with cover coats, low carbon steels for direct-on enamel application, pickling of steel for cover coat application as well as ground coat enameling, pyrolytic oven development, processing of steel for fishscale resistance, and low temperature enamels [1400 ^{O} F]. Additionally, electroluminescent enamels systems were developed, "Nucerite" glass-ceramic coatings for chemical vessels, and water heater coatings. The enameled coatings on steel for water heaters greatly improved the performance of these systems.

Figure 1 – Water heater tanks on a process line.

1965 - 1975
The next decade saw the development of continuous cleaning [catalytic] coatings, wet electrostatic spray application, aluminum enamel spall resistance testing, high

temperature coating technology for steels and alloys for aircraft components, initial development of continuously cast steel, ceramic coated exhaust systems, phosphorescent enamels, dishwasher coatings along with the development of the current dishwasher configuration, vignetting or "highlighting" of appliances [harvest gold, avocado, Coppertone], coil coating of porcelain enamel, initial use of robots for enamel application, electrophoretic enameling of appliance parts and electrostatic powder development which greatly reduced the necessary labor for enamel application. This decade also saw the first significant "energy crisis" starting in 1973 which lasted for over a decade. Near the end of the decade, IF (interstitial free) steel was introduced and refrigerator liners were generally no longer enameled.

Figure 2 – Continuous cleaning oven, interior coating with high surface area and high concentrations of metal oxide in the enamel surfaces.

1975 – 1985

The continuation of the energy crisis stimulated the development of various schemes for solar energy utilization. Enamels were shown to be effective collection surfaces on steel substrates. This decade saw the commercialization of electrostatic powder application of ground coat, cover coats and the "two coat – one fire" systems that are now widely used for appliances. At the same time, the competing technology of electrophorectic application was being utilized commercially. Electronic microcircuits and washing machine components were produced using this technology.

In the late 1970's, the "no nickel-no pickle" ground coat systems were introduced and established commercially. This was stimulated in part by a "nickel shortage" starting in 1969. Cobalt oxide was used in higher quantities to compensate. A "cobalt shortage" in the late 1970's stimulated the development of high nickel ground coats. Electrolytic metal cleaning was evaluated as well as frit "recycling" to minimize wastage in plants. Interest in lower temperature enamels, again a consequence of the energy crisis resulted in development of new enamel systems. A "shortage" of natural gas resulted in many enameling furnaces being converted to electricity. Ceramic fiber lined furnaces were gradually introduced and now predominate in the industry.

<u>1985 to 1995</u>

Early in this decade, work on no-cleanig enameling was experimented with resulting in some success for washing machine outer baskets. No-pickle continuous cleaning [catalytic] enamels were also developed. Continuously cast steels for enameling continued to be further developed such as enameling iron replacement steel. Premilled enamels gained in usage as new enameling plants were designed without traditional mill rooms for milling the enamels. Mixing systems for these "ready-to-use" enamels required significantly less space and labor.

Figure 3 – Various enameled articles. Range, barbeque, and stove components coated by dry powder electrostatic process.

<u>1995 to 2005</u>

The most recent decade has seen further development of the electrostatic and electrophoretic processes. The powder systems are now the norm worldwide for efficient enamel application. Cast iron grates for ranges are being coated electrophoretically. Newer technologies for single coat non-stick coatings, application of sol-gel technology, enameling of newer aluminized steels for cookware and architectural applications, and metallic appearance coatings are becoming commercially developed.

Figure 4 – Commercially produced range with metallic appearance coating on top, door and lower drawer.

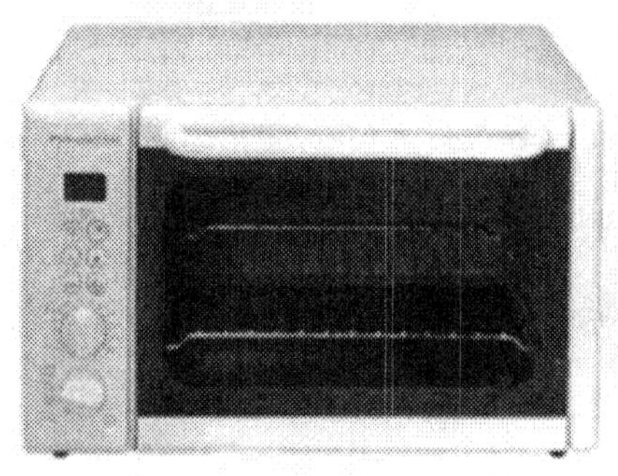

Figure 5 – New non-stick surfaces produced by one coat application of non-stick RealEAse™ coating.

Toaster Oven

Cast Iron Skillet with non-stick RealEase™ coating.

RealEase is a trademark of Ferro Corporation

Other Developments

In addition to the technological developments affecting the enameling industry, significant changes have occurred regarding governmental regulations. Significant efforts were made in the early 1960's to start to control air pollution and industrial waste. About 1974, the Environmental Protection Administration (EPA) began enforcing new regulations for all types of wastes. Additional wastewater regulations came about in the early 1980's and has continued to be tightened. The Occupations Safety and Health Administration (OSHA) became an important part of plant operations in the early 1970's and 1980's. In the early 1990's, ISO became important for manufacturers to achieve high quality operations. In the early part of the new millennium, the Sarbanes-Oxley law greatly increased the need for public companies to adhere to specific operational guidelines.

In the mid-1990's, ISO 1400 and QS 9000 were introduced to further improve environmental stewardship and quality of operations apart from governmental direction. A variety of new methods of work were also introduced: Evolutionary Operation (EVOP), Taguchi problem solving methodology, reintroductions of Statistical Process Control (SPC) and Six Sigma methodologies for continuous improvement of operations. All of these have required development of new skills and streamlining of operations to be effectively utilized. Much of his has been realized with the introduction and continuously evolving microelectronics and communication systems.

Globally, there have been and continue to be significant changes in the industry regarding manufacturing locations for minimal costs. Lower wage countries are now attracting producers to outsource their products in ever greater numbers.

Future

Continued shifts in the industrial production will affect the industry. Raw materials and production are now not limited geographically. Delivery of high quality products to almost any point on the globe is more typical than unusual. The degree of communications today is many orders of magnitude greater than one or two decades ago. The basic evolution and development of technology creating new products has been associated with the manufacturing capability of various societies.

In the April 2005 American Ceramic Society Bulletin, it noted that the Task Force on the Future of American Innovation in its report "Innovate America", the number of students enrolling in science and engineering studies fell 10% between 1994 and 2001. The number of postdoctoral positions at U.S. schools is held by foreign-born scholars, the number of scientific papers published internationally with U. S. authors fell from 38% to 31%. Apart from the aging of the science and engineering workforce, funding for physical science basic research is continuing to decline [www.futureofinnovation.org].[1]

We need to keep alert to new opportunities for adaptation of the basic technology to new products. An example would be enameled substrates[2] for use a heating elements in domestic ranges. Additionally, simplified coatings and processes will enhance the usefulness of porcelain enamels along with intelligent design of components to be enameled.

References

1 "Innovate America", *American Ceramic Society Bulletin,* Vol. 84, No. 4, April 2005.

2. Sridharan, S., Brown, O., Mason, K., Dijkstra, P., Steinbruck, H., Svanbom, P, Ringler, S, "Thick Film Heated Oven With Low Energy Consumption", *Proceedings of The Porcelain Enamel Institute Forum,* Vol. 67, 2005.

THICK FILM HEATED OVEN WITH LOW ENERGY CONSUMPTION

Helga Steinbrück[1], Dr. Srinivasan Sridharan[2, a], Petter Svanbom[3], Orville Brown[4], Dr. Sandrine Ringler[5], Keith Mason[6], and Pieter Dijkstra[7]

[1, 3, 5] Electrolux Major Appliances, Primary Development Cooking, Rothenburg o.d.T, Germany
[2] Ferro Corporation, Posnick Center for Innovative Technology, Independence, OH, USA
[4]Ferro Corporation, Electronic Materials Systems, South Plainfield, NJ, USA
[6]Ferro Corporation, Electronic Materials Systems, Vista, CA, USA
[7]Ferro Corporation, Industrial Coatings Group, Rotterdam, Holland

Abstract
In this work we have demonstrated the concept of a thick film heated oven, and its materials system, that pass the electrical safety requirements for stationary class I heating appliances as outlined in IEC 60335-1. This was accomplished, in part, through the development of low leakage fast-fire enamel dielectrics, compatible resistor & over glaze materials by Ferro Corporation. This was combined with the development of a new thick film heating pattern design from Electrolux Major Appliances. The resulting oven a) exhibits a uniform temperature distribution across the heated surface, b) reaches the required maximum operating temperature of 300°C at less than 1500W per plate, c) has a leakage current of <0.75mA per applied kW at 300°C, d) has a breakdown voltage strength greater than 1250V AC (at >300°C), e) exhibits an excellent life time of >2000 ON-hours without failure, and, f) shows cooking results comparable to conventional ovens with tubular heating elements.

Introduction
A film-heated oven, wherein the cavity walls are directly heated by thick film heaters embedded on the enamel coating to minimize the energy consumption, has been the subject of interest for the appliance industry as a whole, and Electrolux in particular. The main obstacle was achieving the desired electrical characteristics at an acceptable operating temperature of the oven without cracking in the enamel layer.

These problems arise mainly from the different materials involved and the designed shape of the oven. The materials combination, normally, comprises a thin gage low carbon steel substrate, enamels that provide electrical insulation, a resistive heating film that generates heat, and then finally, an overglaze layer to protect the resistor traces. First, conventional porcelain enamels with excessive amounts of alkali oxides in their compositions become electrically conductive for all practical purposes at low temperatures, compared to the oven operating temperatures of about 200 to 350°C. Secondly, due to different thermal and elastic properties of the metal substrate and glass based enamel systems, thermal residual stresses develop, which can lead to cracking when these tensions are not controlled properly.

To resolve the first problem, Ferro developed new alkali free enamels, compatible resistors and over glaze materials that fire in conventional porcelain enamel fast fire furnaces. To resolve the second problem, Electrolux investigated fabricating and optimising (for uniform heat distribution) a Thick Film Heater (TFH for short) on flat steel plates. Then, the electrical and cooking performance of an oven cavity design that incorporates these TFH panels was studied. This work is presented here.

Low leakage enamels

The enamel layer in a film heated oven panel electrically separates the base (low carbon steel) metal substrate and the resistor layer. Therefore, the required electrical performance of low leakage current (LC for short), and high breakdown voltage (BDV for short) are governed by the electrical properties of the enamel layer. These include insulation resistance (IR for short), dielectric properties such as dielectric constant (K), loss factor, as well as thickness of, pore size and pore distribution in the enamel layer. Further, their variations with temperature determine the useful upper operating temperature of the resulting film heated oven panel.

Conventional porcelain enamels for low carbon steel substrates are, in general, based on alkali borosilicate glasses. These glasses fire at a typical peak temperature of 780-850°C

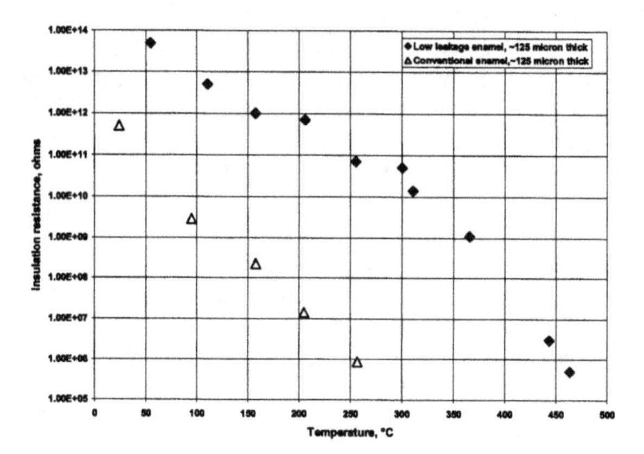

Fig. 1: Comparison of insulation resistance of a) a conventional porcelain enamel, with that of b) a low leakage enamel of the current development

(about 5 minutes above 700°C) in a continuous fast belt furnace wherein the parts are 'in' and 'out' of the furnace in about 20 minutes. Due to excessive alkali ions in these enamels their insulation resistance quickly degrades with rise in temperature, especially in the temperature regime 200-300°C as shown in Fig. 1 (see conventional enamel). In response to

that, Ferro developed a series of alkali free glasses for enamel applications that fire in a typical porcelain enamel furnace firing conditions [1]. As shown in Fig.1, IR of these low leakage enamels is about four to five orders of magnitude higher than that of the conventional enamels in 200-350°C temperature region. Further these electrostatically coated low leakage enamels exhibit a leakage current to power ratio of about 0.12 mA/kW or less at 300°C, for a typical enamel fired thickness of 200-250 μm [1].

Oven cavity panels, electrostatic spray process exhibited often a) uncontrolled cracking in the enamel, and b) premature BDV failures. The former was traced to enamel thickness variation as well as curvatures in the embossed cavity panels. The occurrence of premature BDV failures was traced to bigger bubbles in the enamel which were difficult to control due to the nature of the electrostatic spray process. This premature BDV failure problem gets especially severe, when the enamel thickness is reduced to reduce the warpage of the panel. To overcome these problems, alternate techniques of enamel application were undertaken. The focus was mainly on the screen-printing method for enamel application, to control the enamel thickness variation and to reduce the overall enamel thickness. Further, the oven panel design was changed to flat configuration. Enamel compositions were modified to control: a) bubble formation due to reactions with the metal substrate, b) bubble size, and c) resistor/enamel interactions.

Screen Printing Pastes

Dielectrics (enamels): Two new dielectric (enamel) screen-printing pastes, HT018 and HT051, were developed. As shown in Fig. 2, HT018 is the bonding dielectric that promotes

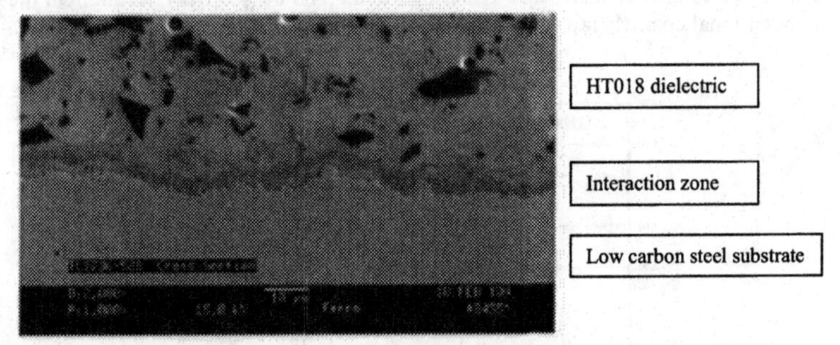

'ig. 2: Cross section of HT018 dielectrics on the low carbon steel substrate fired at 775°C. Vote the formation of almost bubble free interaction zone at the dielectrics/metal interface

xcellent bonding and adhesion to the low carbon steel substrate, without generating ignificant amount of bubbles at the interface. HT051 is the top dielectric. HT051 was esigned to have excellent compatibility with both the bonding dielectric HT018 and the post ired (@630°C) resistor paste HR81-.010 (from Ferro). Further, HT051 is also designed to ave a better thermal expansion matching to the low carbon steel to reduce warpage, which ecomes significant as the dielectric layer thickness increases. Although HT051 dielectric is ated to be the top dielectric, it could also be used as the bonding dielectric in place of IT018.

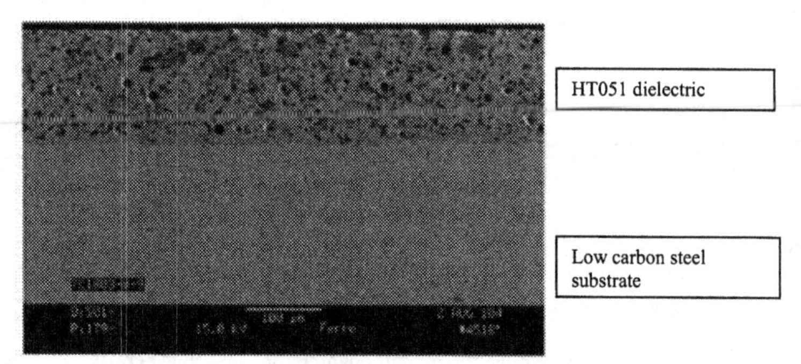

Fig. 3: Cross section of HT051 dielectric on the low carbon steel substrate fired at 780°C. Note the formation of interaction zone without any bubbles formed at the interface. Note also the excellent microstructure of dielectric with very few but fine bubbles

Fig. 3 is the cross section of HT051 dielectric printed and then fired at 780°C, which shows very few but fine bubbles, and good bonding to the low carbon steel substrate. These microstructures passed the critical requirement of BDV >1250V AC for 60 sec at both 25°C and 350°C, without failure for the dielectric thickness of about 125 μm. As we see in Fig. 4, these dielectrics exhibit very low leakage up to 400~450°C and breakdown occurs only at temperatures greater than ~550°C during 1250V AC 60 sec BDV testing. On the other hand, conventional enamels fail this test even at room temperature.

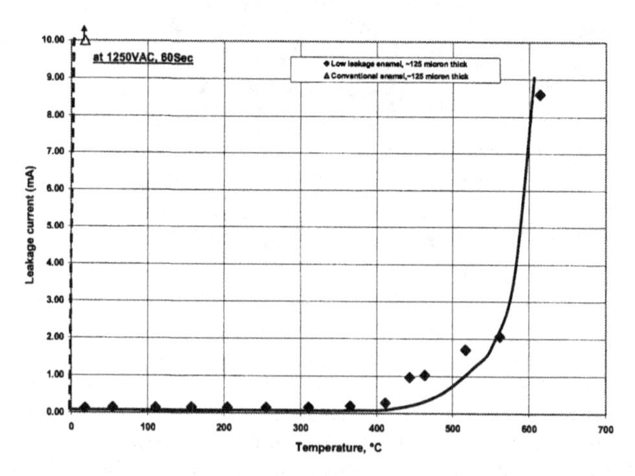

Fig. 4: Observed leakage current as a function of temperature for a) conventional enamel, and b) low leakage dielectrics (enamels), when 1250V AC was applied for 60 sec across the dielectric. The limit of leakage was set 10 mA in the Hi Pot tester

Resistor: The lead free and cadmium free resistor paste HR81-.010 (from Ferro) used has the following characteristics:

- PTC behaviour
- Sheet resistance: 10 mΩ/sq at room temperature
- TCR: 3100 ppm/°C
- Firing temperature: 630-700°C (10 min)

Overglaze: The lead free and cadmium free overglaze paste HT039 (from Ferro) was used for this work [2]. HT039 is designed to minimize any resistance shift due to any overglaze/resistor interaction either during overglaze firing (610~650°C) or during subsequent life time testing. Fig. 5 shows the excellent compatibility between the dielectric/resistor/overglaze materials combination used for this work.

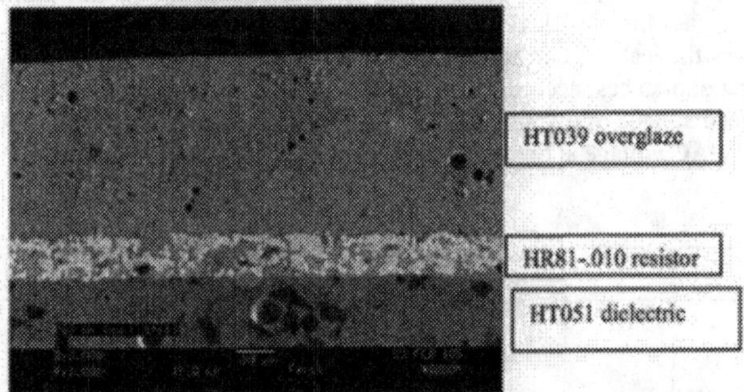

Fig. 5: Cross section of HT051 dielectric/HR81-.010 resistor/HT039 overglaze fired on the low carbon steel substrate showing the excellent compatibility between these three materials

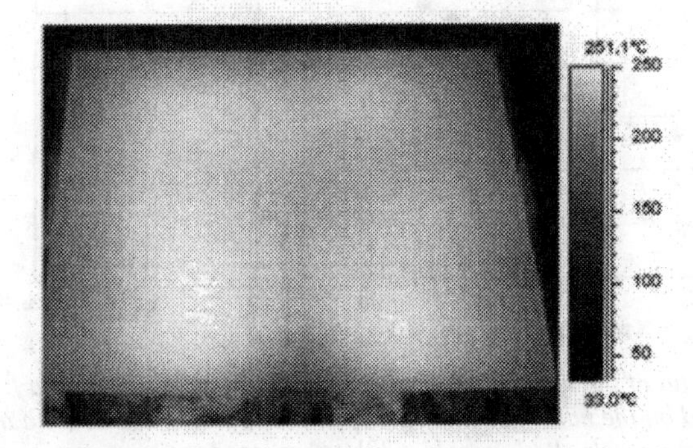

Fig. 6: Observed temperature distribution on the TFH plate in real time IR monitoring experiment

Generation of the film heating pattern

A proprietary heater pattern was developed by Electrolux to get an optimized heat distribution over the whole panel's surface of 426 mm x 386 mm. The temperature distribution due to this pattern at a given operating temperature was simulated for 400V AC application. Then the parameters of the heater design were optimized to minimize the temperature gradient over the entire surface. Once simulated the identified pattern is applied and fired on the substrate, using the resistor paste HR81-.010. The temperature distribution in these real panels was then monitored with an infra red (IR) camera and compared to that of the simulation. In general, simulation predicts the actual performance very well. Fig. 6 shows one such observed temperature distribution on a TFH plate.

Testing

Electrical Safety: The main achievement in this project is the fact that the thick film heating panels pass the electrical safety requirements for electrical household appliances: i.e. the panels pass the leakage current and the high voltage breakdown safety tests for stationary class I heating appliances, according to IEC 60335-1[3]. That means that the leakage current does not exceed a value of 0.75mA per applied kW and no breakdown occurs during 60 sec of applied 1250V AC voltage at operating temperature (~300°C) [4].

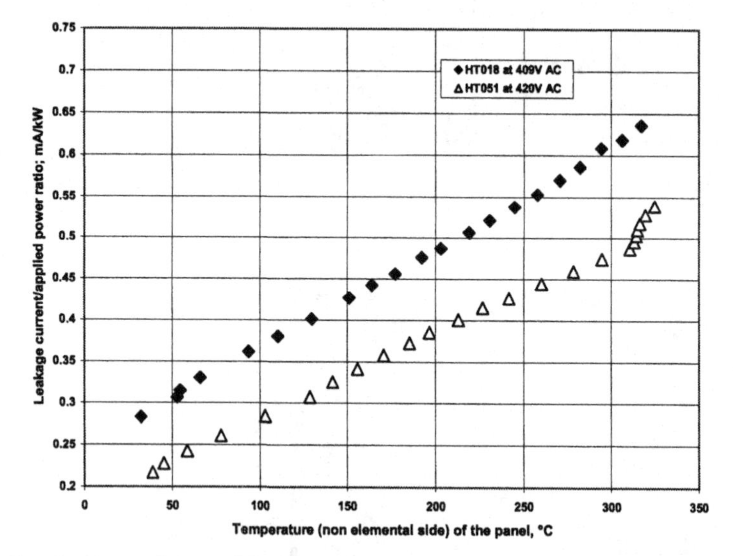

Fig. 7: Observed ratio of the leakage current to applied power (mA/kW) as a function of the temperature reached on the non elemental side of the TFH panels coated with a) HT018 and b) HT051 dielectric enamels. Both panels exhibited a steady state power output of ~1500 watts.

Fig. 7 shows the observed leakage current to applied power ratio as a function of temperature of (non elemental side of) the panel for TFH panels coated with either HT018 or HT051 dielectric as the bond coat. This figure clearly shows that the ratio of the leakage current to applied power for both dielectric enamels, is well within the safety limit of 0.75 mA per applied kW. Even though both dielectric enamels are suitable for passing the leakage current test, HT051 shows a rather increased performance, especially when applied as first layer on the substrate. According to the standard, during the testing the applied voltage has to be set at ~7% higher than under normal conditions. The leakage current is then measured until the device reaches the steady state conditions (which is at around 340°C). As can be seen in Fig. 7, the leakage current is somewhat lower for the panel coated with the H051 dielectric enamel, and hence better performance.

Furthermore, with the present thickness of the dielectric enamel (~200μm), irrespective of the dielectric enamel used (HT018 or HT051), the thick film heating panels pass the high voltage break down test at operating temperatures, when the dielectric enamels are applied in particle free environment.

Heat up rate: Fig. 8 shows that the thick film heated panels reach the steady state operating temperature of 300~320°C in about 8 ~10 minutes.

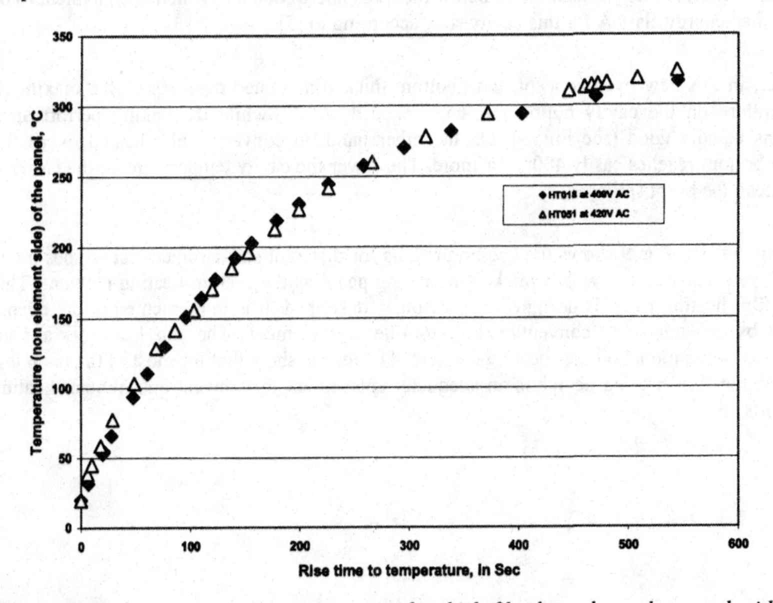

Fig. 8: Plot of time to rise to temperature for thick film heated panels coated with two different dielectric enamels.

Lifetime test: The endurance test for the film heating panels is performed according to an Electrolux internal verification standard for tubular heating elements. The test requires that minimum eight panels shall be run at operating temperature while cycling 45 min ON and 15 min OFF. The test goes on until 2000 ON-hours are reached. No failure may occur during this test period.

Nine prototypes of thick film heating panels were put on lifetime test (after having passed the electrical safety tests). Out of these two prototypes failed due to reasons other than that in lifetime testing. So the failures were not truly representative of life time testing. The remaining seven panels passed the lifetime testing. The change in resistance of these seven panels did not exceed ±5% compared to the starting value at room temperature. Thus these panels passed the lifetime test.

Thick film heated oven

Implementation: Electrolux has developed a concept to implement the thick film heating panel as a bottom heating element in the oven. The cavity bottom is heated directly by the thick film heating panel. With this method of implementation, the prototype oven has got a low stay with its energy consumption below the threshold of 800Wh, which is equivalent to or better than energy class A for this cavity size, according to [5].

Further, in this new oven concept, with bottom thick film heated oven panel, the maximum temperature on the cavity bottom does not exceed 300°C, while the baking performance remains equally good (see Fig. 9). On the other hand, in conventionally heated ovens the cavity bottom reaches easily 400°C or more. The lower the cavity temperature is, the easier it is to clean the burnt spillages, etc.

Baking results: Fig. 9 shows the cooking results for different standardized test recipes made in an oven that was run with a thick film heating panel as the bottom-heating element. This thick film heating panel is designed as previously described. It is implemented in the oven's cavity by replacing the conventional bottom-heating element. The top heater is a non-modified conventional tubular heating element. The results show that for most of the main test recipes, the film heating device is an adequate replacement for conventional tubular heating elements.

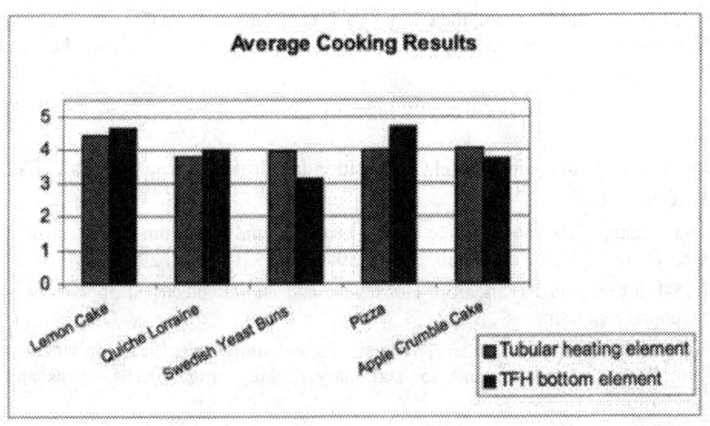

Fig. 9: Cooking results for the standard recipes obtained with a) conventional oven with tubular heating element, and b) novel oven with thick film heating panel at the bottom. The results are rated from 0 to 5 where 5 is the maximum score.

Conclusions

A working concept of thick film heated (TFH) household oven with low energy consumption has been demonstrated in this work.

This oven:

a) Fulfils the electrical safety and long-term reliability requirements for stationary heating appliances
b) With bottom TFH implementation meets the energy standards of class A ovens
c) Provides good to very good baking results

Reliable and reproducible prototyping of these thick film heating panels was done using screen printing as the method of applications for enamel, resistor and overglaze materials. The thick film material system is lead free and cadmium free and is RoHS compliant

Acknowledgements

The authors want to thank:

- the managements of Electrolux Major Appliances, Rothenburg o.d.T, Germany and Ferro Corporation, Cleveland, OH, USA for their encouragements in publishing this work
- Richard Wiesinger, manager of the enamel shop, at Electrolux, in Rothenburg, Germany, for new and inspiring ideas, the whole enamel workshop for the firing tests, and Mr. Reiner Horstmann of Electrolux for the various laboratory electrical testing , and
- Mr. Dave Gnizak, Microscopy Specialist, of Ferro Corporation for SEM work; Dr. Dick Abrams, of Ferro Corporation, for critically reading the manuscript; Mr. Fred

Neuhaus & Mr. Achadu Unogwu, the Glass Lab Technicians at Ferro Corporation, and Sandy Schallers, Thick film Inks Technician, at Ferro Corporation, for their help in carrying out experiments.

References

[1] S. Sridharan, et.al., "Porcelain Enamel Composition for Electronic Applications", U.S. Patent 5,998,037, Dec 7, (1999)

[2] S. Sridharan, et.al., "Electronic Device having Lead free and Cadmium free Electronic Overglaze Applied Thereof", U.S. Patent Pub. No. US2004/0018931A1, Jan 29, (2004)

[3] IEC 60335-1 International Standard, "Household and similar electrical appliances – Safety – Part 1: General requirements", (2001)

[4] IEC 60335-2-6 International Standard, "Household and similar electrical appliances – Safety – Part 2-6: Particular requirements for stationary cooking ranges, hobs, ovens and similar appliances", (2002)

[5] EN 50304 European Standard, "Electric ovens for household use – methods for measuring the energy consumption", (2001)

9 781574 98278